Birkhäuser

Advances in Mechanics and Mathematics

Advances in Continuum Mechanics

Volume 53

This series includes collective thematic books and monographs on advanced research topics in continuum mechanics. It is a unique forum for reference reviews discussing critically foundational issues in classical field theories under the light of up-to-date problems raised by the offspring of condensed matter physics and industrial needs, including those concerning the mechanics of microstructures and the relevant multi-field (or phase-field) representations. The series aims also to show new mathematical methods in continuum mechanics and/or recent developments of existing powerful practice. Also, the series pays attention to applications with concrete conceptual content. The series editors select editors of books on specific themes, which may split into sequential volumes.

All manuscripts are peer-reviewed to meet the highest standards of scientific literature:

- Series editors provide their final decision for inclusion in the series after manuscripts have been fully peer-reviewed;
- For contributed volumes, volume editors conduct the anonymous peer-review process before final submission;
- The peer-review process is anonymous and confidential in order to ensure reviewers provide impartial opinions

Brian Straughan

Thermal Convection with a Cattaneo Flux Law

Birkhäuser

Brian Straughan
Department of Mathematical Sciences
Durham University
Durham, UK

ISSN 1571-8689 ISSN 1876-9896 (electronic)
Advances in Mechanics and Mathematics
ISSN 2524-4639 ISSN 2524-4647 (electronic)
Advances in Continuum Mechanics
ISBN 978-3-031-80884-5 ISBN 978-3-031-80885-2 (eBook)
https://doi.org/10.1007/978-3-031-80885-2

This book is published under the imprint Birkhäuser, www.birkhauser-science.com by the registered company Springer Nature Switzerland AG
The registered company address is: Gewerbestrasse 11, 6330 Cham, Switzerland

Preface

This book presents an account of thermal convection in a layer of fluid heated from below when Fourier's law is not assumed. Instead, we focus on an evolutionary equation for the heat flux known as Cattaneo's heat flux equation. Several other effects are incorporated including Guyer–Krumhansl terms, the Mariano effect, Kelvin–Voigt fluid theory, and higher gradient fluid theories. We believe this is the first account of such a class of problem in book form.

I was introduced to Cattaneo's equation by the late and great mathematical physicist Professor Dario Graffi of the University of Bologna in 1979. He also introduced me to his then Doctoral student Franca Franchi who had just finished her thesis on uniqueness, thermodynamics, and wave motion in problems involving Cattaneo's equation. I am indebted to Dario Graffi and Franca Franchi for many illuminating discussions on the topic of Cattaneo heat flow theory. I have also benefitted greatly through many discussions on this topic with Mauro Fabrizio, Maurizio Gentile, Pedro Jordan, and Angelo Morro. It is a pleasure to thank Paolo Mariano for his encouragement to publish this book, and to thank Chris Eder for help with editorial matters. It is also a pleasure to thank Paula Francis for copyediting the manuscript.

This work was supported in part by a Leverhulme Trust Emeritus Fellowship, grant number EM-2019-022/9.

Durham,
December 2023

Brian Straughan

Contents

Chapter 1
Introduction

1.1 Cattaneo's equation and heat waves in a rigid conductor

First sound is the classical mechanism by means of which we hear, i.e. a disturbance of pressure (or density) that propagates through a continuous medium such as air or water. Second sound is a more recent phenomenon involving the propagation of heat as a temperature wave. A seminal paper which derived a theoretical model to describe the propagation of heat as a wave is Cattaneo (1948). This book analyses various applications of the equation proposed by Cattaneo (1948), and its subsequent ramifications, to problems in the field of thermal convection.

The classical theory of heat propagation is via diffusion where a temperature field diffuses through a continuous body, and the theory is usually described by an energy balance law coupled with a constitutive equation for the heat flux. The usual approach employs Fourier's law where the heat flux at time t is assumed to be proportional to the temperature gradient at the same time. Cattaneo (1948) generalized this and introduced the idea of a relaxation time into the relationship between the heat flux and the temperature gradient. In the late 1960's and early 1970's Physicists performed experiments which did indeed show that a thermal disturbance could travel as a wave and this has acted as an impulse to much subsequent theoretical work in this area. As (Straughan, 2011a, pages 1-3) observes thermal pulse propagation has been experimentally observed under accurate conditions in solid helium (He^3 and He^4), in sodium fluoride, in bismuth, in sodium iodide and in lithium fluoride, and details may be found in that book, with an up to date analysis of some of these experiments given by Kovács and Ván (2018). This aspect of second sound is a low temperature phenomenon, the experiments just cited having been performed in the cryogenic 1-20°K range.

In addition to work in a rigid body, further experiments (see details in (Straughan, 2011a, pages 1-3)) showed that a thermal wave could also propagate in an elastic body. These experiments employed a very pure crystal of sodium fluoride and evaporated manganin heaters and lead detectors onto opposing faces of the crystal and were thereby able to transmit heat pulses through their sample. These careful

B. Straughan, *Thermal Convection with a Cattaneo Flux Law*, Advances in Mechanics and Mathematics 53, https://doi.org/10.1007/978-3-031-80885-2_1

experiments revealed the existence of three distinct waves. There was a longitudinal elastic wave which travels fastest, a transverse elastic wave, and also a thermal wave. When the temperature was below 8°K, three distinct waves were observed, the fastest being the longitudinal one, the transverse one next fastest, while the thermal wave was slowest.

Within fluid mechanics, the effect of temperature upon wave propagation of a disturbance has a long history. A brief review of relevant historical aspects is given in (Straughan, 2011a, page 2).

Work involving Cattaneo's equation is usually referred to as employing Maxwell–Cattaneo theory since relaxation ideas were also present in earlier work of Maxwell (1867). King et al. (1998) explain the connection with Maxwell, writing Cattaneo used arguments ... *based upon results in the kinetic theory of gases which go back to the original work of* (*Maxwell,* 1867, *equation* (143)). *Maxwell's results concerning* Cattaneo's equation ... *have been rigorously derived and extended in the treatise of Truesdell and Muncaster* (1980).

It is important to emphasize that Maxwell–Cattaneo theory for heat flow is a very rich area in modern research. In addition to heat flow, the idea behind Maxwell–Cattaneo theory where a relaxation effect is included for an appropriate flux is being employed in many diverse research areas. Such models do not generally suffer from the effect of the Fourier law of infinite speed of heat propagation, and allow heat to travel with a finite wavespeed. The technique behind Maxwell–Cattaneo theory whereby a Fourier like flux equation is substituted with a Cattaneo like equation for the flux has been applied in many mundane areas. For example, in nanoscale mechanics, Sellitto et al. (2020), Jou et al. (2010), Vadasz et al. (2005), Ruggeri and Sugiyama (2021), Elshehabey et al. (2024); in thermal waves with a radiation source, Zampoli and Jordan (2022); in temperature wave motion, Jordan and Lambers (2021a,b), Carillo and Jordan (2020), Christov and Jordan (2005), Jordan (2015), Christov (2016, 2019); in convection in stellar atmospheres, Govender and Govinder (2004), Falcón (1998), Falcón and Labrador (2001), Herrera and Falcón (1995b,c,a), Herrera and Santos (1997), Herrera and Martínez (1998), Herrera (2019); in collapse in stellar structures, Govender and Thirukkanesh (2009), Govender et al. (2012), Herrera and Santos (2004); in destruction of tumours, Andres and Pinnau (2019, 2022), Blauth et al. (2019); in drug delivery models, Ferreira and de Oliveira (2010); in describing how vegetation forms into patterns on landscapes, Currò and Valenti (2021), Consolo et al. (2017, 2019, 2020), Consolo et al. (2022), Grifò et al. (2023); in chemotaxis, Barbera and Valenti (2018), Dolak and Hillen (2003), Wang and Hillen (2008); in pollution, Barbera et al. (2010), Gentile and Straughan (2016); in biological tissues, Mitra et al. (1995), Vedavarz et al. (1992); in the drying of sand, Meyer (2006); in cryovulcanology in one of Saturn's moons, Bargmann et al. (2008); in phase changes, Miranville and Quintanilla (2009), Liu et al. (2009); in nuclear reactor technology, Espinosa-Paredes and Espinosa-Martinez (2009); in skin burns, Dai et al. (2008); in analysing chronic wasting disease, Barbera and Pollino (2023); in the medical technique of radiofrequency heating, López Molina et al. (2008); in the spread of a virus, Barbera et al. (2008); in traffic flow, Jordan (2005); in energy storage, Elshehabey et al. (2024); in diffusion with chemical reactions,

Al Ghoul (2004); in gene expression time delay, Alfifi (2022); potentially to re-entry of a satellite into the atmosphere of a planet, Ruggeri and Sugiyama (2021); and to the control of fish stocks, Niwa (1998); and this technique is important as a surgical procedure in the elimination of cardiac arrhythmias, tumours, in heating of the cornea, or in the treatment of gastroesophageal reflux disease. Further examples may be found in the book by Straughan (2011a).

There are also important applications of extensions of Cattaneo's equation, for example when dissipation of the heat flux is included, yielding Guyer–Krumhansl theory. Incorporation of Guyer–Krumhansl theory is pertinent since Ván et al. (2017), Fülöp et al. (2018), suggest that this theory may be more relevant than Fourier theory, even at room temperatures, see also Berezovski (2016, 2019), Capriz et al. (2021), Carlomagno et al. (2021), Cimmelli (2021), Famà et al. (2021), Rogolino and Cimmelli (2021). This book looks at a variety of issues connected with convection using Cattaneo theory and, in particular, we do include accounts of the contemporary issues just mentioned.

1.2 Notation, definitions

Standard indicial notation is used throughout this book together with the Einstein summation convention for repeated indices. Standard vector or tensor notation is also employed where appropriate. For example, we write

$$u_x \equiv \frac{\partial u}{\partial x} \equiv u_{,x} \qquad u_{i,t} \equiv \frac{\partial u_i}{\partial t},$$

$$u_{i,i} \equiv \frac{\partial u_i}{\partial x_i} \equiv \sum_{i=1}^{3} \frac{\partial u_i}{\partial x_i} = \frac{\partial u}{\partial x} + \frac{\partial v}{\partial y} + \frac{\partial w}{\partial z},$$

$$u_j u_{i,j} \equiv u_j \frac{\partial u_i}{\partial x_j} \equiv \sum_{j=1}^{3} u_j \frac{\partial u_i}{\partial x_j}, \qquad i = 1, 2 \text{ or } 3,$$

where we have also written $\mathbf{u} = (u, v, w)$. In the interests of clarity, if we write out the last term, $u_j u_{i,j}$, in full, then the three expressions are

$$u\frac{\partial u}{\partial x} + v\frac{\partial u}{\partial y} + w\frac{\partial w}{\partial z},$$
$$u\frac{\partial v}{\partial x} + v\frac{\partial v}{\partial y} + w\frac{\partial v}{\partial z},$$
$$u\frac{\partial w}{\partial x} + v\frac{\partial w}{\partial y} + w\frac{\partial w}{\partial z},$$

corresponding to $i = 1, 2$ or 3. In the case where a repeated index sums over a range different from 1 to 3 this will be pointed out in the text. Note that

$$u_j u_{i,j} \equiv (\mathbf{u} \cdot \nabla)\mathbf{u} \qquad \text{and} \qquad u_{i,i} \equiv \operatorname{div} \mathbf{u}.$$

As indicated above, a subscript t denotes partial differentiation with respect to time. When a superposed dot is used it means the material derivative, eg.

$$\dot{u}_i \equiv \frac{\partial u_i}{\partial t} + u_j \frac{\partial u_i}{\partial x_j},$$

or

$$\dot{T} = \frac{\partial T}{\partial t} + u_i \frac{\partial T}{\partial x_i},$$

where u_i in the equations above is the velocity field and $T(\mathbf{x}, t)$ is another field such as temperature.

To further clarify the indicial notation used, we note for example, the product of two second order tensors A_{ij} and B_{ij}, say $\mathbf{C} = \mathbf{AB}$, may be written

$$\begin{aligned} C_{ij} &= A_{ik} B_{kj} \\ &\equiv \sum_{k=1}^{3} A_{ik} B_{kj} \,. \end{aligned}$$

Further, if $\mathbf{x}$ and $\mathbf{b}$ are vectors then $\mathbf{Ax} = \mathbf{b}$ may be written as

$$\begin{aligned} b_i &= A_{ik} x_k \\ &\equiv \sum_{k=1}^{3} A_{ik} x_k \,. \end{aligned}$$

We now derive some expressions which are useful for later. Recall the definitions of the Kronecker delta, δ_{ij}, and the alternating tensor, ϵ_{ijk}. Namely,

$$\delta_{ij} = \begin{cases} 1, & \text{if } i = j; \\ 0, & \text{if } i \neq j, \end{cases}$$

i.e. $\delta_{11} = \delta_{22} = \delta_{33} = 1$, the rest are zero. Also,

$$\begin{aligned} \epsilon_{123} = \epsilon_{231} &= \epsilon_{312} = +1, \\ \epsilon_{213} = \epsilon_{321} &= \epsilon_{132} = -1, \end{aligned}$$

the rest are zero.

In terms of the Kronecker delta and the alternating tensor,

$$\epsilon_{ijk} \epsilon_{irs} = \delta_{jr} \delta_{ks} - \delta_{js} \delta_{kr}$$

and then for a vector field $\mathbf{A}$,

$$(\operatorname{curl} \mathbf{A})_i = \epsilon_{ijk} A_{k,j}$$

and

$$\begin{aligned}(\text{curl curl } \mathbf{A})_i =& \epsilon_{ijk}\epsilon_{krs}A_{s,rj} \\ =& \epsilon_{kij}\epsilon_{krs}A_{s,rj} \\ =& (\delta_{ir}\delta_{js} - \delta_{is}\delta_{jr})A_{s,rj} \\ =& A_{j,ij} - \Delta A_i \,. \end{aligned} \tag{1.1}$$

Thus, if v_i is a divergence free vector field, $v_{i,i} = 0$, and then

$$(\text{curl curl } \mathbf{v})_i = -\Delta v_i \,,$$

where $\Delta = \partial^2/\partial x^2 + \partial^2/\partial y^2 + \partial^2/\partial z^2$ is the Laplace operator.

The vector product (or cross product) of two vectors **A** and **B** may be written in component form in indicial notation as

$$(\mathbf{A} \times \mathbf{B})_i = \epsilon_{ijk}A_jB_k \,.$$

If we let $\mathbf{k} = (0, 0, 1)$ then a particular vector product which occurs frequently in analysis of thermal convection in a rotating frame is $\mathbf{u} \times \mathbf{k}$ which may be written in component form as

$$\begin{aligned}(\mathbf{u} \times \mathbf{k})_i =& \epsilon_{ijk}u_jk_k \\ =& \epsilon_{ijk}u_j\delta_{k3} \\ =& \epsilon_{ij3}u_j \,. \end{aligned}$$

Another frequently used expression is curl $(\mathbf{u} \times \mathbf{k})$ and this is

$$\begin{aligned}\big[\text{curl } (\mathbf{u} \times \mathbf{k})\big]_i =& \epsilon_{irs}(\mathbf{u} \times \mathbf{k})_{s,r} \\ =& \epsilon_{irs}\epsilon_{sab}(u_ak_b)_{,r} \\ =& \epsilon_{sir}\epsilon_{sab}k_bu_{a,r} \\ =& (\delta_{ia}\delta_{rb} - \delta_{ib}\delta_{ra})\delta_{b3}u_{a,r} \\ =& (\delta_{ia}\delta_{r3} - \delta_{i3}\delta_{ra})u_{a,r} \\ =& u_{i,3} - \delta_{i3}u_{a,a} \end{aligned}$$

and so if div $\mathbf{u} = 0$, i.e. if $u_{i,i} = 0$, then

$$\big[\text{curl } (\mathbf{u} \times \mathbf{k})\big]_i = u_{i,3} \,. \tag{1.2}$$

Let us put $\mathbf{u} = (u, v, w)$ and $\mathbf{x} = (x, y, z)$. We then observe that the third component of expression (1.2) yields,

$$\big[\text{curl } (\mathbf{u} \times \mathbf{k})\big]_3 = \frac{\partial w}{\partial z} \,. \tag{1.3}$$

When dealing with anisotropic media we have recourse to deal with diagonal matrices and so we introduce the notation

$$\mathbf{A} = \mathrm{diag}\,(a_1, a_2, a_3) \tag{1.4}$$

to mean $\mathbf{A}$ is the diagonal matrix

$$\mathbf{A} = \begin{pmatrix} a_1 & 0 & 0 \\ 0 & a_2 & 0 \\ 0 & 0 & a_3 \end{pmatrix}.$$

The letter Ω will denote a fixed, bounded region of 3-space with boundary, Γ, sufficiently smooth to allow applications of the divergence theorem. When we are dealing with convection problems we handle motion in a plane layer, say $\{(x, y) \in \mathbb{R}^2\} \times \{z \in (0, d)\}$. In this case, we usually refer to functions that have an (x, y) behaviour which is repetitive in the (x, y) direction, such as regular hexagons. The periodic cell defined by such a shape and its Cartesian product with $(0, d)$ will be denoted by V. The boundary of the period cell V will be denoted by ∂V. Such a hexagonal convection cell is depicted in figure 3.1 in section 3.4.

The symbols $\|\cdot\|$ and $(\cdot,\cdot)$ will denote, respectively, the L^2 norm on Ω or V, and the inner product on $L^2(\Omega)$ or $L^2(V)$, where the context will define whether Ω or V is to be used, e.g.,

$$\int_V f^2 dV = \|f\|^2 \quad \text{and} \quad (f, g) = \int_V fg\, dV,$$

with equivalent definitions for Ω. We sometimes have recourse to use the norm on $L^p(\Omega)$, $1 < p < \infty$, and then we write

$$\|f\|_p = \left(\int_\Omega |f|^p dx\right)^{1/p}.$$

We introduce the ideas of stability and instability in the context of the following system of reaction-diffusion equations,

$$\begin{aligned} \frac{\partial u}{\partial t} &= \frac{\partial}{\partial x_j}\left(a_{ij}\frac{\partial u}{\partial x_i}\right) + f(u, v), \\ \frac{\partial v}{\partial t} &= \frac{\partial}{\partial x_j}\left(b_{ij}\frac{\partial v}{\partial x_i}\right) + g(u, v), \end{aligned} \tag{1.5}$$

where f and g are known nonlinear functions and where $\mathbf{x} \in \Omega \subset \mathbb{R}^3$. (In practice equations (1.5) would be defined with suitable boundary conditions.) The coefficients $a_{ij} = a_{ij}(\mathbf{x})$ and $b_{ij} = b_{ij}(\mathbf{x})$ are symmetric and positive-definite in the sense that $a_{ij} = a_{ji}$, $b_{ij} = b_{ji}$ and

$$a_{ij}\xi_i\xi_j \geq a_0\xi_i\xi_i, \qquad b_{ij}\xi_i\xi_j \geq b_0\xi_i\xi_i, \qquad \forall\xi_i,$$

and for constants $a_0 > 0$ and $b_0 > 0$.

We introduce notation in the context of a steady solution to (1.5), namely a solution $\bar{u}, \bar{v}$ satisfying

$$\begin{aligned} 0 &= \frac{\partial}{\partial x_j}\Big(a_{ij}\frac{\partial \bar{u}}{\partial x_i}\Big) + f(\bar{u}, \bar{v}), \\ 0 &= \frac{\partial}{\partial x_j}\Big(b_{ij}\frac{\partial \bar{v}}{\partial x_i}\Big) + g(\bar{u}, \bar{v}). \end{aligned} \tag{1.6}$$

(We could equally deal with the stability of a time-dependent solution, but many of the problems encountered in this book are for stationary solutions and at this juncture it is as well to keep the ideas as simple as possible.)

Let (w, r) be a perturbation to the solution to equation (1.6), i.e. put $u = \bar{u} + w(\mathbf{x}, t)$, $v = \bar{v} + r(\mathbf{x}, t)$. Then, it is seen from (1.5) and (1.6) that (w, r) satisfies the system

$$\begin{aligned} \frac{\partial w}{\partial t} - \frac{\partial}{\partial x_j}\Big(a_{ij}\frac{\partial (w+\bar{u})}{\partial x_i}\Big) - \big[f(\bar{u}+w, \bar{v}+r) - f(\bar{u}, \bar{v})\big] = 0, \\ \frac{\partial r}{\partial t} - \frac{\partial}{\partial x_j}\Big(b_{ij}\frac{\partial (r+\bar{v})}{\partial x_i}\Big) - \big[g(\bar{u}+w, \bar{v}+r) - g(\bar{u}, \bar{v})\big] = 0. \end{aligned} \tag{1.7}$$

To discuss the concept of linearized instability we linearize (1.7) which means we keep only the terms which are linear in w and r. From a Taylor series expansion of f and g we have

$$\begin{aligned} f(\bar{u}+w, \bar{v}+r) =& f(\bar{u}, \bar{v}) + w\frac{\partial f}{\partial u}(\bar{u}, \bar{v}) + r\frac{\partial f}{\partial v}(\bar{u}, \bar{v}) \\ &+ O(w^2, wr, r^2), \\ g(\bar{u}+w, \bar{v}+r) =& g(\bar{u}, \bar{v}) + w\frac{\partial g}{\partial u}(\bar{u}, \bar{v}) + r\frac{\partial g}{\partial v}(\bar{u}, \bar{v}) \\ &+ O(w^2, wr, r^2). \end{aligned} \tag{1.8}$$

Expansions (1.8) are inserted into (1.7) and then with the aid of equations (1.6) we may derive the linearized equations satisfied by w and r, namely

$$\begin{aligned} \frac{\partial w}{\partial t} - \frac{\partial}{\partial x_j}\Big(a_{ij}\frac{\partial w}{\partial x_i}\Big) - wf_u(\bar{u}, \bar{v}) - rf_v(\bar{u}, \bar{v}) = 0, \\ \frac{\partial r}{\partial t} - \frac{\partial}{\partial x_j}\Big(b_{ij}\frac{\partial r}{\partial x_i}\Big) - wg_u(\bar{u}, \bar{v}) - rg_v(\bar{u}, \bar{v}) = 0, \end{aligned} \tag{1.9}$$

where $f_u = \partial f/\partial u$, etc.

Since (1.9) is a system of linear equations we may introduce an exponential time dependence in (w, r) so that $w = e^{\sigma t}s(\mathbf{x})$ and $r = e^{\sigma t}q(\mathbf{x})$. Then (1.9) yields

$$\begin{aligned} \sigma s - \frac{\partial}{\partial x_j}\Big(a_{ij}\frac{\partial s}{\partial x_i}\Big) - sf_u(\bar{u}, \bar{v}) - qf_v(\bar{u}, \bar{v}) = 0, \\ \sigma q - \frac{\partial}{\partial x_j}\Big(b_{ij}\frac{\partial q}{\partial x_i}\Big) - sg_u(\bar{u}, \bar{v}) - qg_v(\bar{u}, \bar{v}) = 0. \end{aligned} \tag{1.10}$$

We say that the steady solution $(\bar{u}, \bar{v})$ to (1.6) is *linearly unstable* if

$$Re(\sigma) > 0,$$

where $Re(\sigma)$ denotes the real part of σ. Equations (1.10) (together with appropriate boundary conditions) constitute an eigenvalue problem for σ. For many of the problems discussed in this book the eigenvalues may be ordered so that

$$Re(\sigma_1) > Re(\sigma_2) > \dots$$

For linear instability we then need only ensure $Re(\sigma_1) > 0$.

We now introduce the concept of nonlinear stability. Let $w_0(\mathbf{x}) = w(\mathbf{x}, 0)$, $r_0(\mathbf{x}) = r(\mathbf{x}, 0)$, be the initial data functions associated to the solution (w, r) of equations (1.7). The steady solution $(\bar{u}, \bar{v})$ to (1.6) is *nonlinearly asymptotically stable* if and only if for each $\epsilon > 0$ there is a $\delta = \delta(\epsilon)$ such that

$$\|w_0\|^2 + \|r_0\|^2 < \delta \Rightarrow \|w(t)\|^2 + \|r(t)\|^2 < \epsilon, \qquad \forall t > 0, \tag{1.11}$$

and there exists γ with $0 < \gamma \leq \infty$ such that

$$\|w_0\|^2 + \|r_0\|^2 < \gamma \Rightarrow \lim_{t\to\infty} \left(\|w(t)\|^2 + \|r(t)\|^2\right) = 0. \tag{1.12}$$

If $\gamma = \infty$, we say the solution is *unconditionally* nonlinearly stable (or simply refer to it as being asymptotically stable), otherwise for $\gamma < \infty$ the solution is *conditionally* (nonlinearly) stable. For nonlinear stability problems it is an important goal to derive parameter regions for unconditional nonlinear stability, or at least conditional stability with a finite initial data threshold (i.e. finite, non-vanishing, radius of attraction). It is important to realise that the linearization as in (1.9) and (1.10) can only yield linear *instability*. It tells us nothing whatsoever about stability. There are many equations for which nonlinear solutions will become unstable well before the linear instability analysis predicts this, cf. Lu and Shao (2003). Also, when an analysis is performed with $\gamma < \infty$ in (1.12) this yields conditional nonlinear stability, i.e. nonlinear stability for only a restricted class of initial data.

We have only defined stability with respect to the $L^2(\Omega)$ norm in (1.11) and (1.12). However, sometimes it is convenient to use an analogous definition with respect to some other norm or positive-definite solution measure. It will be clear in the text when this is the case. When we refer to continuous dependence on the initial data we mean a phenomenon like (1.11). Thus, a solution (w, r) to equations (1.7) depends continuously on the initial data if a chain of inequalities like (1.11) holds.

Throughout the book we make frequent use of inequalities. In particular, we often use the Cauchy–Schwarz inequality for two functions f and g, i.e.

$$\int_\Omega fg\,dx \leq \left(\int_\Omega f^2 dx\right)^{1/2} \left(\int_\Omega g^2 dx\right)^{1/2}, \tag{1.13}$$

or what is the same in L^2 norm and inner product notation,

$$(f,g) \leq ||f|| \, ||g||. \tag{1.14}$$

Hölder's inequality is

$$(f,g) \leq ||f||_p \, ||g||_q , \tag{1.15}$$

where

$$\frac{1}{p} + \frac{1}{q} = 1.$$

Writing in terms of integrals this is

$$\int_\Omega fg \, dx \leq \left(\int_\Omega |f|^p dx \right)^{1/p} \left(\int_\Omega |g|^q dx \right)^{1/q} . \tag{1.16}$$

For Ω a domain in $\mathbb{R}^3$, with a function u which is zero on the boundary Γ, the Poincaré inequality is

$$\lambda_1 ||\mathbf{u}||^2 \leq ||\nabla u||^2 , \tag{1.17}$$

where $\lambda_1(\Omega) > 0$ is the first eigenvalue in the membrane problem for Ω. This can be alternatively written as

$$\lambda_1 \int_\Omega u^2 \, dx \leq \int_\Omega |\nabla u|^2 dx. \tag{1.18}$$

For Ω a domain in $\mathbb{R}^3$, with a function u which is zero on the boundary Γ, one has the Sobolev inequality, (Gilbard and Trudinger, 1977, page 148),

$$||u||_6 \leq C||\nabla u||, \tag{1.19}$$

for an absolute constant C. This may be equivalently written

$$\left(\int_\Omega u^6 \, dx \right)^{1/3} \leq C^2 \int_\Omega |\nabla u|^2 \, dx . \tag{1.20}$$

For the same class of functions u, another useful relation, which is also refereed to as a Sobolev inequality is

$$||u||_4 \leq c_3(\Omega)||\nabla u||, \tag{1.21}$$

where c_3 is a positive constant depending on the geometry of Ω. This can be rewritten

$$\int_\Omega u^4 \, dx \leq c_3^4 \left(\int_\Omega |\nabla u|^2 \, dx \right)^2 . \tag{1.22}$$

An estimate for the constant in (1.22) may be found using the Cauchy–Schwarz inequality, (1.17) and (1.19) as

$$\int_\Omega u^4\,dx \le \Big(\int_\Omega u^2\,dx\Big)^{1/2}\Big(\int_\Omega u^6\,dx\Big)^{1/2}$$
$$\le \frac{C^3}{\sqrt{\lambda_1}}\Big(\int_\Omega |\nabla u|^2\,dx\Big)^2 .$$

For a vector function $\mathbf{u}$ which is zero on the boundary Γ there are two further Sobolev inequalities which we require. For Ω a domain in $\mathbb{R}^2$ one has for a constant $c_1 > 0$,

$$\|\mathbf{u}\|_4^4 \le c_1^2\|\mathbf{u}\|^2\|\nabla\mathbf{u}\|^2. \tag{1.23}$$

For Ω a domain in $\mathbb{R}^3$ one has for a constant $c_2 > 0$,

$$\|\mathbf{u}\|_4^4 \le c_2^2\|\mathbf{u}\|\|\nabla\mathbf{u}\|^3 . \tag{1.24}$$

Details of these inequalities and their constants may be found in (Payne, 1964, pages 132–135).

The arithmetic-geometric mean inequality (with a constant weight $\alpha > 0$) is, for $a, b \in \mathbb{R}$,

$$ab \le \frac{1}{2\alpha}a^2 + \frac{\alpha}{2}b^2, \tag{1.25}$$

and this is easily seen to hold since

$$\Big(\frac{a}{\sqrt{\alpha}} - \sqrt{\alpha}b\Big)^2 \ge 0.$$

Another inequality we frequently have recourse to is Young's inequality, which for $a, b \in \mathbb{R}$ we may write as

$$ab \le \frac{|a|^p}{p} + \frac{|b|^q}{q}, \qquad \frac{1}{p} + \frac{1}{q} = 1, \quad p, q \ge 1. \tag{1.26}$$

1.3 Maxwell–Cattaneo theory

Let T denote the temperature of a rigid body at time t and spatial position $\mathbf{x}$. The paper by Cattaneo (1948) is one which has had a major influence on virtually every paper on thermal waves. Cattaneo (1948) begins with the classical diffusion equation for heat, for example in one space dimension,

$$\frac{\partial T}{\partial t} = D\frac{\partial^2 T}{\partial x^2}, \qquad x \in \mathbb{R}, t > 0, \tag{1.27}$$

with the initial data

$$T(x,0) = f(x), \qquad x \in \mathbb{R}. \tag{1.28}$$

Equation (1.27) together with (1.28) may be solved by a Fourier transform method, see e.g. Sneddon (1995). Cattaneo (1948) observes that the solution is given by

$$T(x,t) = \frac{1}{2\sqrt{\pi D t}} \int_{-\infty}^{\infty} f(\xi) \exp\left[\frac{-(x-\xi)^2}{4Dt}\right] d\xi. \tag{1.29}$$

He further observes that the solution (1.29) essentially has an infinite speed of propagation, for example, f might be 0 outside a finite set but (1.29) implies $\theta \neq 0$ $\forall x \in \mathbb{R}$ for $t > 0$.

If T denotes the temperature of a rigid solid, ρ its density, c its specific heat, and k its thermal conductivity, then Cattaneo (1948) notes that equation (1.27) arises from the energy balance law

$$\rho c \frac{\partial T}{\partial t} = -\frac{\partial q}{\partial x}, \tag{1.30}$$

together with the Fourier law of heat conduction,

$$q = -k \frac{\partial T}{\partial x}. \tag{1.31}$$

In equations (1.30) and (1.31), q is the heat flux. In equation (1.27) we would take $D = k/\rho c$. In order to obtain a finite speed of propagation, Cattaneo (1948) employs a very interesting argument essentially based on statistical mechanics for a gas. He argues that q in (1.31) may be replaced by the relation

$$q = -k \frac{\partial T}{\partial x} + \sigma \frac{\partial^2 T}{\partial x \partial t}. \tag{1.32}$$

The coefficient σ is given in equation (17) of Cattaneo (1948). To derive a further relation Cattaneo (1948) differentiates (1.32) with respect to t to obtain

$$\frac{\partial q}{\partial t} = -k \frac{\partial^2 T}{\partial x \partial t} + \sigma \frac{\partial^3 T}{\partial x \partial t^2}$$

and then he forms the relation

$$\sigma \frac{\partial q}{\partial t} + kq = -k^2 \frac{\partial T}{\partial x} + \sigma^2 \frac{\partial^3 T}{\partial x \partial t^2}. \tag{1.33}$$

To bring this into line with modern terminology we put $\tau = \sigma/k (> 0)$ and write as

$$\tau \frac{\partial q}{\partial t} + q = -k \frac{\partial T}{\partial x} + \tau\sigma \frac{\partial^3 T}{\partial x \partial t^2}. \tag{1.34}$$

The coefficient τ may be calculated from equations (12), (14) and (17) of Cattaneo (1948) and is a ratio of statistical mechanical averages over molecular velocities and positions.

At this point Cattaneo (1948) argues that one may discard the last term in (1.34) due to its smallness by comparison to the other terms. However, if we retain it the resulting system of equations is (1.30) and (1.34) which may be combined to yield the equation for the temperature

$$\tau\frac{\partial^2 T}{\partial t^2}+\frac{\partial T}{\partial t}=\frac{k}{\rho c}\frac{\partial^2 T}{\partial x^2}-\frac{\sigma^2}{\rho ck}\frac{\partial^4 T}{\partial x^2\partial t^2}\,. \tag{1.35}$$

To return to the mainstream argument of Cattaneo (1948), on page 93 he argues that the last term in equation (1.34)may be discarded. His famous system of equations is then derived from equations (1.30) and (1.34) as

$$\begin{aligned}\rho c\frac{\partial T}{\partial t}&=-\frac{\partial q}{\partial x}\,,\\ \tau\frac{\partial q}{\partial t}+q&=-k\frac{\partial T}{\partial x}\,.\end{aligned} \tag{1.36}$$

Cattaneo (1948) observes that eliminating q leads to the damped wave equation

$$\tau\frac{\partial^2 T}{\partial t^2}+\frac{\partial T}{\partial t}=\frac{k}{\rho c}\frac{\partial^2 T}{\partial x^2}\,. \tag{1.37}$$

A way to derive equation (1.36) which is frequently employed in the literature is to argue to replace the Fourier law (1.31) by a delay equation

$$q(x,t+\tau)=-k\frac{\partial T}{\partial x}(x,t)\,. \tag{1.38}$$

In other words, the heat flux does not depend instantaneously on the temperature gradient at a point; there is a short time lag before the effect is felt. One then expands the left hand side of equation (1.38) using a Taylor series so that

$$q(x,t+\tau)=q(x,t)+\tau\frac{\partial q}{\partial t}(x,t)+O(\tau^2). \tag{1.39}$$

The $O(\tau^2)$ terms are neglected and then employing equation (1.38) one may arrive at equation (1.36). However, care must be taken with this approach since Jordan et al. (2008) show that if one combines the energy balance law $(1.36)_1$ together with equation (1.38) with no approximation then the resulting delay equation displays a lack of continuous dependence on the initial data, i.e. an instability. (For interest, we point out that Quintanilla (2008) and Quintanilla and Jordan (2009) show that well posedness may be recovered if one combines the equation (1.38) with a two-temperature energy balance equation.)

The Cattaneo system (1.36) and its three-dimensional equivalent,

$$\begin{aligned}\rho c\frac{\partial T}{\partial t}&=-\frac{\partial q_i}{\partial x_i}\,,\\ \tau\frac{\partial q_i}{\partial t}+q_i&=-k\frac{\partial T}{\partial x_i}\,,\end{aligned} \tag{1.40}$$

have been the subject of immense study in the literature.

For a general theory the thermal relaxation coefficient τ will depend on temperature as discussed in (Straughan, 2011a, pages 13,14). In this book we mostly analyse situations where τ is constant. However, a recent interesting analysis and discussion of the Maxwell-Cattaneo equation with a temperature dependent thermal relaxation term, τ, is given by Carillo and Jordan (2020).

1.3.1 Graffi's work on relaxation theory in electromagnetism

We commence with a very brief description of work of Graffi (1936) in a generalized theory of Maxwell's equations in electromagnetism which involves memory terms, cf. Fabrizio (2011), Franchi and Straughan (2023). The point of this is to observe that Graffi (1936) derived a theory for the electric field which is essentially a generalization of the form (1.40).

Graffi (1936) proposes a generalization of Maxwell's equations which involve memory terms. In fact, in the medium outside an antenna, Graffi (1936) begins with Faraday's law

$$\frac{\partial \mathbf{B}}{\partial t} = -\mathrm{curl}\, \mathbf{E}, \tag{1.41}$$

where $\mathbf{E}$ is the electric field and $\mathbf{B}$ is the magnetic induction. He also writes Ampère's law with the electric displacement correction of Maxwell, namely

$$\mathrm{curl}\, \mathbf{H} = \frac{\partial \mathbf{D}}{\partial t} + \mathbf{j}, \tag{1.42}$$

where $\mathbf{H}$ is the magnetic field, $\mathbf{j}$ is the current density, and $\mathbf{D}$ is the electric displacement current.

For the current density Graffi (1936) proposes

$$\mathbf{j} = \sigma \mathbf{E} + \int_0^t \beta(t-s)\mathbf{E}(s)ds, \tag{1.43}$$

where σ is the electrical conductivity and the β term represents the history dependence of the current on the electric field. Graffi (1936) notes that certain materials display dielectric hysteresis and so proposes the dependence of $\mathbf{D}$ upon $\mathbf{E}$ as

$$\mathbf{D} = \epsilon \mathbf{E} + \int_0^t \gamma(t,s)\mathbf{E}(s)ds. \tag{1.44}$$

Here ϵ is the usual electric permittivity and the γ term represents the history dependence on the electric field. It is worth observing that a basis for equations like (1.44) and the analogous one involving $\mathbf{B}$ and $\mathbf{H}$ is discussed in detail in a thermodynamic context in the books of Fabrizio and Morro (2003) and Morro and Giorgi (2023).

Graffi (1936) observes that defining the functions λ and α by

$$\lambda(t) = \gamma(t,t) + \sigma \quad \text{and} \quad \alpha = \gamma'(t,s) + \beta(t-s)$$

then equations (1.42) - (1.44) lead to the equation

$$\operatorname{curl} \mathbf{H} = \epsilon \frac{\partial \mathbf{E}}{\partial t} + \lambda \mathbf{E} + \int_0^t \alpha(t,s)\mathbf{E}(s)ds. \tag{1.45}$$

Furthermore, if $\mathbf{B} = \mu\mathbf{H}$ where μ is the magnetic permeability, then equation (1.41) becomes

$$\operatorname{curl} \mathbf{E} = -\mu \frac{\partial \mathbf{H}}{\partial t}. \tag{1.46}$$

Equations (1.45) and (1.46) are Graffi's system for the electric and magnetic fields $\mathbf{E}$ and $\mathbf{H}$.

We observe that if $\alpha \equiv 0$, i.e. if $\gamma'(t,s) = -\beta(t-s)$, then (1.45) is very like the Cattaneo law, equation $(1.40)_2$. To see that $\alpha \equiv 0$ is not meaningless we may select the realistic and frequently made choices

$$\gamma(t,s) = \gamma_0\, e^{-\delta(t-s)}, \qquad \beta = \beta_0\, e^{-\delta(t-s)},$$

which lead to

$$\alpha = \gamma' + \beta = (\beta - \gamma_0\delta)\, e^{-\delta(t-s)},$$

and so when $\beta = \gamma_0\delta$ then $\alpha \equiv 0$. A system of equations which is analogous to Cattaneo's system follows directly from (1.41) and (1.42) with $\mathbf{D} = \epsilon\mathbf{E}$ and $\mathbf{j} = \sigma\mathbf{E}$, as shown in section 9.2.

1.4 Related models

There have been many recent very interesting articles dealing with Maxwell–Cattaneo theories, but also proposing and analysing generalizations of this theory. We cite for example, the works of Ván et al. (2017), Ván and Fülöp (2012), Kovács and Ván (2015), Sellitto et al. (2020), Mariano (2017), Mariano et al. (2022), Capriz et al. (2021), Dell'Oro and Pata (2023), Christov (2007), Nika (2023), Nika and Muntean (2022), Cimmelli et al. (2010), Fülöp et al. (2018), Morro (2018), Fabrizio et al. (2017), Giorgi et al. (2024), Straughan (2011a). In this section we review some of these ideas.

Each theory we examine begins with an energy balance equation of form

$$\rho c_p \frac{\partial T}{\partial t} = -\frac{\partial q_i}{\partial x_i} \tag{1.47}$$

where ρ is the density of the body, c_p is the specific heat (at constant pressure), T is the temperature, and q_i is the heat flux. To this is appended an equation yielding the heat flux in some form. For classical heat conduction we have Fourier's law

$$q_i = -k\frac{\partial T}{\partial x_i}, \tag{1.48}$$

where k is the thermal conductivity of the material. This results in the system

$$\begin{aligned} \rho c_p \frac{\partial T}{\partial t} &= -\frac{\partial q_i}{\partial x_i}, \\ q_i &= -k\frac{\partial T}{\partial x_i}. \end{aligned} \tag{1.49}$$

Eliminating q_i leads to the well known parabolic heat equation

$$\frac{\partial T}{\partial t} = \kappa \Delta T, \tag{1.50}$$

where $\kappa = k/\rho c_p$ is the thermal diffusivity, providing k is constant.

For completeness we include the Maxwell–Cattaneo system

$$\begin{aligned} \rho c_p \frac{\partial T}{\partial t} &= -\frac{\partial q_i}{\partial x_i}, \\ \tau \frac{\partial T}{\partial t} + q_i &= -k\frac{\partial T}{\partial x_i}, \end{aligned} \tag{1.51}$$

where τ is the relaxation time. Eliminating q_i leads to the following damped wave equation

$$\tau \frac{\partial^2 T}{\partial t^2} + \frac{\partial T}{\partial t} = \kappa \Delta T. \tag{1.52}$$

A commonly used model which allows for dissipation of the heat flux is the Guyer–Krumhansl system, lucidly described by Ván et al. (2017) and this has form

$$\begin{aligned} \rho c_p \frac{\partial T}{\partial t} &= -\frac{\partial q_i}{\partial x_i}, \\ \tau \frac{\partial q_i}{\partial t} + q_i &= -k\frac{\partial T}{\partial x_i} + a_1 \frac{\partial^2 q_j}{\partial x_j \partial x_i} + a_2 \Delta q_i, \end{aligned} \tag{1.53}$$

where a_1, a_2 are positive constants. If one eliminates the heat flux then the following strongly damped wave equation is obtained for T,

$$\tau \frac{\partial^2 T}{\partial t^2} + \frac{\partial T}{\partial t} - \frac{(a_1 + a_2)}{\rho c_p} \Delta \frac{\partial T}{\partial t} = \kappa \Delta T. \tag{1.54}$$

The Jeffries system, see Ván and Fülöp (2012), Kovács and Ván (2015), Mariano (2017) is

$$\begin{aligned} \rho c_p \frac{\partial T}{\partial t} &= -\frac{\partial q_i}{\partial x_i}, \\ \tau \frac{\partial q_i}{\partial t} + q_i &= -k\frac{\partial T}{\partial x_i} + a_3 \frac{\partial^2 T}{\partial t \partial x_i}. \end{aligned} \tag{1.55}$$

Upon elimination of the heat flux one obtains the temperature equation

$$\tau \frac{\partial^2 T}{\partial t^2} + \frac{\partial T}{\partial t} + \frac{a_3}{\rho c_p} \Delta \frac{\partial T}{\partial t} = \kappa \Delta T. \tag{1.56}$$

If $a_3 < 0$ this is essentially the same as (1.54).

A system attributed to Albert Green and Paul Naghdi, see Ván and Fülöp (2012), Kovács and Ván (2015), Mariano (2017) is

$$\begin{aligned} \rho c_p \frac{\partial T}{\partial t} &= -\frac{\partial q_i}{\partial x_i}, \\ \tau \frac{\partial q_i}{\partial t} &= -k \frac{\partial T}{\partial x_i} + a_2 \Delta q_i , \end{aligned} \tag{1.57}$$

Upon elimination of the heat flux in this case one obtains the following temperature equation

$$\tau \frac{\partial^2 T}{\partial t^2} - a_2 \Delta \frac{\partial T}{\partial t} = \kappa \Delta T. \tag{1.58}$$

Note that the form of damping in (1.58) is different from that in (1.54).

A generalization of the above Green–Naghdi system is given by Sellitto et al. (2013), Fülöp et al. (2018) as

$$\begin{aligned} \rho c_p \frac{\partial T}{\partial t} &= -\frac{\partial q_i}{\partial x_i}, \\ \tau \frac{\partial q_i}{\partial t} + q_i &= -k \frac{\partial T}{\partial x_i} + a_2 \Delta q_i , \end{aligned} \tag{1.59}$$

Upon elimination of the heat flux one now obtains the following temperature equation

$$\tau \frac{\partial^2 T}{\partial t^2} + \frac{\partial T}{\partial t} - a_2 \Delta \frac{\partial T}{\partial t} = \kappa \Delta T. \tag{1.60}$$

The Graffi–Fabrizio theory of heat conduction replaces $\mathbf{E}$ by $\mathbf{q}$ and curl $\mathbf{H}$ by $-k\nabla T$ in equation (1.45). One then replaces (1.46) with the energy balance equation. This yields the Graffi - Fabrizio system of equations

$$\begin{aligned} \rho c_p \frac{\partial T}{\partial t} &= -\frac{\partial q_i}{\partial x_i}, \\ \tau q_{i,t} + q_i + \int_0^t \alpha(t,s) q_i(\mathbf{x}, s)\, ds &= -k \nabla T. \end{aligned} \tag{1.61}$$

With $\alpha \equiv 0$ equations (1.61) reduce to the Maxwell–Cattaneo system. Upon elimination of q_i one derives the following equation for the temperature field $T(\mathbf{x}, t)$,

$$\tau \frac{\partial^2 T}{\partial t^2} + \frac{\partial T}{\partial t} + \int_0^t \alpha(t,s) \frac{\partial T}{\partial s}\, ds = \frac{k}{\rho c} \Delta T. \tag{1.62}$$

With the choice $\alpha = \alpha_0 \, e^{-\omega(t-s)}$ one sees that T satisfies the equation

$$\tau \frac{\partial^3 T}{\partial t^3} + (1 + \omega\tau) \frac{\partial^2 T}{\partial t^2} + (\alpha_0 + \omega) \frac{\partial T}{\partial t} = \frac{k}{\rho c} \Delta \frac{\partial T}{\partial t} + \frac{\omega k}{\rho c} \Delta T. \tag{1.63}$$

There is much interest in heat transfer at the nanoscale and with this in mind we include a system of equations of Christov (2007), who argues that when length scales are important one should include higher gradients of temperature in the constitutive theory. This theory does not contain a Cattaneo effect but the relevant system of equations is

$$\begin{aligned} \rho c_p \frac{\partial T}{\partial t} &= -\frac{\partial q_i}{\partial x_i}, \\ q_i &= -k \frac{\partial T}{\partial x_i} + k_1 \Delta \frac{\partial T}{\partial x_i}, \end{aligned} \tag{1.64}$$

where $k_1 > 0$. Elimination of q_i in this case shows the temperature satisfies the partial differential equation

$$\frac{\partial T}{\partial t} = \kappa \Delta T - \kappa_1 \Delta^2 T, \tag{1.65}$$

where $\kappa_1 = k/\rho c_p$. It should be noted that a constitutive theory like $(1.64)_2$ is rigorously justified using homogenization theory by Nika and Muntean (2022), Nika (2023). Christov (2007) also considers a variation of $(1.64)_2$ which has form

$$-\chi \Delta q_i + q_i = -k \frac{\partial T}{\partial x_i} + k_1 \Delta \frac{\partial T}{\partial x_i}.$$

This, together with $(1.64)_1$ leads to the temperature equation

$$\frac{\partial T}{\partial t} - \chi \Delta \frac{\partial T}{\partial t} = \kappa \Delta T - \kappa_1 \Delta^2 T. \tag{1.66}$$

One can combine the theory of Christov (2007) with the Cattaneo (1948) model and then deal with the Christov–Cattaneo system of heat conduction

$$\begin{aligned} \rho c_p \frac{\partial T}{\partial t} &= -\frac{\partial q_i}{\partial x_i}, \\ \tau \frac{\partial q_i}{\partial t} + q_i &= -k \frac{\partial T}{\partial x_i} + k_1 \Delta \frac{\partial T}{\partial x_i}. \end{aligned} \tag{1.67}$$

Upon elimination of the heat flux one now derives the temperature equation

$$\tau \frac{\partial^2 T}{\partial t^2} + \frac{\partial T}{\partial t} = \kappa \Delta T - \kappa_1 \Delta^2 T. \tag{1.68}$$

Interestingly Iesan (2023) derives the following equation for the temperature field

$$\tau \frac{\partial^2 T}{\partial t^2} = \kappa \Delta T - \kappa_1 \Delta^2 T\,. \tag{1.69}$$

The work of Iesan (2023) is based on a rigorous continuum thermodynamic theory where higher gradients of strain and temperature are involved in the constitutive theory. Further recent articles in this vein in continuum thermodynamics are those of Fabrizio et al. (2022, 2023). I believe this area of work is likely to become very much studied in the future due to mundane applications to such as biological tissue behaviour, cf. Thompson et al. (2010).

One may develop the heat flux further in a Taylor series expansion in $(1.67)_2$, cf. Quintanilla and Racke (2007), Chirita (2018), Fernández and Quintanilla (2023), Dell'Oro and Pata (2023), to derive a higher order Christov–Cattaneo like system of form

$$\begin{aligned} &\rho c_p \frac{\partial T}{\partial t} = -\frac{\partial q_i}{\partial x_i}, \\ &\tau_1 \frac{\partial^2 q_i}{\partial t^2} + \tau \frac{\partial q_i}{\partial t} + q_i = -k \frac{\partial T}{\partial x_i} + k_1 \Delta \frac{\partial T}{\partial x_i}, \end{aligned} \tag{1.70}$$

where τ_1 is a second relaxation time. Upon elimination of the heat flux the following temperature equation is found

$$\tau_1 \frac{\partial^3 T}{\partial t^3} + \tau \frac{\partial^2 T}{\partial t^2} + \frac{\partial T}{\partial t} = \kappa \Delta T - \kappa_1 \Delta^2 T\,. \tag{1.71}$$

Third order time derivatives are becoming important in some areas of continuum mechanics such as Burgers' fluids, and this and other examples may be found in the interesting recent article of Fernández and Quintanilla (2023).

In addition to the above models which are linear, there are several systems proposed to govern heat flow which involve nonlinear partial differential equations. We outline some of these. A model which directly generalizes the Maxwell–Cattaneo theory is that of Sellitto et al. (2020). Here the system of equations has form

$$\begin{aligned} &\rho c_p \frac{\partial T}{\partial t} = -\frac{\partial q_i}{\partial x_i}, \\ &\tau \frac{\partial q_i}{\partial t} + q_i = -k\Big(1 + \frac{\tau q^2}{\kappa T^2}\Big) \frac{\partial T}{\partial x_i} + \frac{2\tau}{\rho c_p T} q_j \frac{\partial q_j}{\partial x_i}, \end{aligned} \tag{1.72}$$

where $q = |\mathbf{q}|$.

Cimmelli et al. (2010) propose a nonlinear form of Guyer–Krumhansl theory for which the equations are

$$\begin{aligned} &\rho c_p \frac{\partial T}{\partial t} = -\frac{\partial q_i}{\partial x_i}, \\ &\tau \frac{\partial q_i}{\partial t} + q_i = -k \frac{\partial T}{\partial x_i} + a_1 \Delta q_i + a_2 \frac{\partial^2 q_k}{\partial x_k \partial x_i} + a_4 q_k \frac{\partial q_k}{\partial x_i}\,. \end{aligned} \tag{1.73}$$

Morro (2018) also proposes another nonlinear form of Guyer–Krumhansl theory, and the equations are

$$\begin{aligned} \rho c_p \frac{\partial T}{\partial t} &= -\frac{\partial q_i}{\partial x_i}, \\ \tau \frac{\partial q_i}{\partial t} + q_i &= -k \frac{\partial T}{\partial x_i} + a_1 \Delta q_i + a_2 \frac{\partial^2 q_k}{\partial x_k \partial x_i} + a_4 q_k \frac{\partial q_k}{\partial x_i} + a_5 \frac{\partial (q_k q_i)}{\partial x_k}. \end{aligned} \tag{1.74}$$

Finally, we observe that Coleman and Mizel (1963) proposed a non-Fourier model of heat conduction which is not of Cattaneo type. However, the nonlinear constitutive theory is likely to be of interest in thermal convection studies. Their model is based on the equations

$$\begin{aligned} \rho c_p \frac{\partial T}{\partial t} &= -\frac{\partial q_i}{\partial x_i}, \\ q_i &= -k \frac{\partial T}{\partial x_i} - \beta |\nabla T|^2 \frac{\partial T}{\partial x_i} - \gamma \Delta T \frac{\partial T}{\partial x_i} - \delta \frac{\partial T}{\partial x_j} \frac{\partial^2 T}{\partial x_j \partial x_i}, \end{aligned} \tag{1.75}$$

where the coefficients β, γ, δ may be taken to be constants. An analysis of thermal convection using both linear instability theory and nonlinear energy stability theory for a model which has constitutive theory not dissimilar to (1.75) is contained in Straughan (2007).

One could write down a Cattaneo version of (1.75), which would have form

$$\begin{aligned} \rho c_p \frac{\partial T}{\partial t} &= -\frac{\partial q_i}{\partial x_i}, \\ \tau \frac{\partial q_i}{\partial t} + q_i &= -k \frac{\partial T}{\partial x_i} - \beta |\nabla T|^2 \frac{\partial T}{\partial x_i} - \gamma \Delta T \frac{\partial T}{\partial x_i} - \delta \frac{\partial T}{\partial x_j} \frac{\partial^2 T}{\partial x_j \partial x_i}. \end{aligned} \tag{1.76}$$

The thermodynamic and stability properties of solutions to this system may well be of interest.

A class of constitutive theory which has some resemblance to that of $(1.75)_2$ was proposed by Rodrigues and Urbano (1998). Their theory would involve $(1.75)_1$ together with

$$q_i = -k_1 |\nabla T|^{p-2} \frac{\partial T}{\partial x_i}, \tag{1.77}$$

where $1 < p < \infty$. One could clearly extend this and have

$$q_i = -k \frac{\partial T}{\partial x_i} - k_1 |\nabla T|^{p-2} \frac{\partial T}{\partial x_i}. \tag{1.78}$$

A Cattaneo model involving this relation could involve the equations

$$
\begin{aligned}
&\rho c_p \frac{\partial T}{\partial t} = -\frac{\partial q_i}{\partial x_i}, \\
&\tau \frac{\partial q_i}{\partial t} + q_i = -k \frac{\partial T}{\partial x_i} - k_1 |\nabla T|^{p-2} \frac{\partial T}{\partial x_i}.
\end{aligned}
\tag{1.79}
$$

There are some very interesting developments for the propagation of sound which employ Cattaneo like ideas to avoid the infinite signal speed paradox of classical second - order in time strongly damped models of nonlinear acoustics. These result in partial differential equation models such as the Moore–Gibson–Thompson equation, or the Jordan–Moore–Gibson–Thompson equation. We do not explicitly review this field, but there are beautiful analyses of solution behaviour and many references to the growing literature which may be found in Kaltenbacher et al. (2012), Marchand et al. (2012), Kaltenbacher and Nikolić (2019), Racke and Said Houari (2021), Nikolić and Said Houari (2021), Said Houari (2022), Zampoli and Jordan (2024). The Jordan - Moore - Gibson - Thompson equation is, see e.g. Kaltenbacher and Nikolić (2021),

$$
\tau \frac{\partial^3 \psi}{\partial t^3} + \frac{\partial^2 \psi}{\partial t^2} - (\delta + \tau c^2)\Delta \frac{\partial \psi}{\partial t} - c^2 \Delta \psi = \frac{1}{2}\frac{\partial}{\partial t}\Big[\kappa \Big(\frac{\partial \psi}{\partial t}\Big)^2 + \sigma |\nabla \psi|^2\Big].
$$

Here ψ is the acoustic velocity potential and the derivation of this equation relies on using the Maxwell–Cattaneo equation for the heat flux. Within the field of poroacoustics Zampoli and Jordan (2024) derive the following system of equations by using the Cattaneo–Christov equation for the heat flux

$$
\begin{aligned}
&\frac{\partial^2 u}{\partial t^2} + \delta \frac{\partial u}{\partial t} - b_0^2 \frac{\partial^2 u}{\partial x^2} = -b_0^2 \Big(1 + \frac{\sigma_0 T_0}{p_0}\Big) \frac{\partial^2 T}{\partial t \partial x}, \\
&\frac{\partial T}{\partial t} + \tau \frac{\partial^2 T}{\partial t^2} - \tau a_0^2 \frac{\partial^2 T}{\partial x^2} = -(\gamma - 1)\Big(\frac{\partial u}{\partial x} + \tau \frac{\partial^2 u}{\partial t \partial x}\Big),
\end{aligned}
$$

where u and T are the speed and temperature. If one were interested in thermal convection in a compressible fluid saturating a porous medium then the above approach may be relevant.

Chapter 2
Classical thermal convection

2.1 Thermal convection

Before discussing the topic of thermal convection when the heat flux satisfies a Maxwell–Cattaneo law we briefly describe the subject in the context of classical theory where Fourier's law holds. The linear theory of instability in classical thermal convection is covered in great detail in the book by Chandrasekhar (1981). A treatment of nonlinear stability theory for the classical problem employing an energy method may be found in Straughan (2004). However, it is instructive to give brief details here so that the peculiarities which arise when Maxwell–Cattaneo theory is employed are easier to follow.

Thermal convection in a fluid appears to have been first investigated seriously in the 1800s. Wesfreid (2017) notes that Weber (1855) describes cells with convective motion in a thin layer of alcohol. Thomson (1882) also observed similar: in particular, (Larmor and Thomson, 1912, pages 136–141), writing on the work of Thomson (1882) write that, *James Thomson observed a tessellated structure, considerably resembling such patterns as are exhibited by vertical basaltic columns when exposed to view in a horizontal section, ..., in a tub of opaque looking water, in the yard of a roadside inn, ...* They continue to explain how Thomson created an experiment to verify this in soapy water. (Larmor and Thomson, 1912, page 136) remark *This subject has been investigated independently and with extensive developments by Prof. Bénard of Bordeaux in a thesis published in the Annales de Chemie,* 1901 ...: *see also Bénard, Revue Génerale des Sciences,* 1900.

Wesfreid (2017) describes how Bénard performed careful experiments and observed regular hexagonal cells and regular shaped convective rolls, see e.g. Bénard (1900). In fact, the topic of thermal convection is now often referred to as Bénard convection.

Thermal convection is motion in a fluid which is due to a temperature gradient which in turn induces a density gradient which then drives a convective motion in the fluid. The classical example is to consider a fluid contained in the horizontal layer $\{0 < z < d\}$ with the temperatures, T, at the boundaries fixed so that

B. Straughan, *Thermal Convection with a Cattaneo Flux Law*, Advances in Mechanics and Mathematics 53, https://doi.org/10.1007/978-3-031-80885-2_2

$$T = T_U, \quad z = d; \qquad T = T_L, \quad z = 0;$$

where T_U, T_L, are constants with $T_L > T_U$. Since the temperature at the lower boundary is greater than that at the upper boundary this creates a vertical temperature gradient in the layer if gravity is acting in the downward direction. The fluid at the base of the layer becomes hotter, expands, and if the layer is deep enough a convective motion commences. If the fluid is essentially incompressible, then it must descend somewhere else in the layer, and typically a hexagonal structure arises with the fluid rising in the centre of each hexagon and descending at the walls.

To describe thermal convection mathematically we now introduce the Navier–Stokes equations which govern the evolutionary behaviour of a linearly viscous, incompressible, fluid. These are firstly introduced in an isothermal setting, and after that thermal effects are introduced by the addition of an energy balance law for the fluid.

2.2 The Navier–Stokes equations for incompressible fluid flow

The basic equations for isothermal flow of a fluid are the balance law of linear momentum,

$$\rho \dot{v}_i = \sigma_{ji,j} + \rho f_i \,, \tag{2.1}$$

where ρ is the density, $v_i(\mathbf{x}, t)$ is the velocity field, f_i is an externally supplied body force, σ_{ij} is the (symmetric) Cauchy stress tensor, and the superposed dot denotes the material derivative.

For many flows it is sufficient to regard the fluid as incompressible. In this situation, the equation of balance of mass becomes

$$v_{i,i} = 0, \tag{2.2}$$

i.e. the velocity field is solenoidal. For a linearly viscous fluid the Cauchy stress tensor has form

$$\sigma_{ij} = -p\delta_{ij} + 2\mu d_{ij} \,, \tag{2.3}$$

where p is the pressure, $\mu > 0$ is the dynamic viscosity, and d_{ij} is the symmetric part of the velocity gradient, i.e.

$$d_{ij} = \frac{1}{2}(v_{i,j} + v_{j,i}) \,. \tag{2.4}$$

Thus, for isothermal flow equations (2.1) - (2.4) yield the Navier–Stokes equations

$$\begin{aligned} &\frac{\partial v_i}{\partial t} + v_j \frac{\partial v_i}{\partial x_j} = -\frac{1}{\rho}\frac{\partial p}{\partial x_i i} + \nu \Delta v_i + f_i, \\ &\frac{\partial v_i}{\partial x_i i} = 0, \end{aligned} \tag{2.5}$$

where ν (positive constant) is the kinematic viscosity, and Δ is the Laplacian.

We shall study equations (2.5) on a bounded spatial region Ω ($\subset \mathbb{R}^3$) with $\mathbf{v}$ given on the boundary Γ, i.e.,

$$v_i(\mathbf{x},t) = \hat{v}_i(\mathbf{x},t), \qquad \mathbf{x} \in \Gamma. \tag{2.6}$$

In the interests of clarity, and to ensure full understanding of the indicial notation used throughout this book we write out equations (2.5) in full as a system of partial differential equations for the four dependent variables v_1, v_2, v_3 and p. It is convenient to use the notation $\mathbf{v} = (u, v, w)$ and $\mathbf{x} = (x, y, z)$, and then system (2.5) is equivalent to

$$\begin{aligned}
&\frac{\partial u}{\partial t} + u\frac{\partial u}{\partial x} + v\frac{\partial u}{\partial y} + w\frac{\partial u}{\partial z} = -\frac{1}{\rho}\frac{\partial p}{\partial x} + \nu\Delta u + f_1\,,\\
&\frac{\partial v}{\partial t} + u\frac{\partial v}{\partial x} + v\frac{\partial v}{\partial y} + w\frac{\partial v}{\partial z} = -\frac{1}{\rho}\frac{\partial p}{\partial y} + \nu\Delta v + f_2\,,\\
&\frac{\partial w}{\partial t} + u\frac{\partial w}{\partial x} + v\frac{\partial w}{\partial y} + w\frac{\partial w}{\partial z} = -\frac{1}{\rho}\frac{\partial p}{\partial z} + \nu\Delta w + f_3\,,\\
&\frac{\partial u}{\partial x} + \frac{\partial v}{\partial y} + \frac{\partial w}{\partial z} = 0\,.
\end{aligned} \tag{2.7}$$

2.3 The Balance of Energy and the Boussinesq Approximation

When we deal with motion of a fluid driven by buoyancy forces, such as those caused by heating the fluid, equations (2.5) are not sufficient and it is necessary to also add an equation for the temperature field, T. This is the equation for the *balance of energy*,

$$\rho T\dot{\varepsilon} = \sigma_{ij}d_{ij} + \rho r - q_{i,i}, \tag{2.8}$$

where $\varepsilon, \sigma_{ij}, d_{ij}, r$, and q_i are the internal energy, stress tensor, symmetric part of the velocity gradient, external supply of heat, and heat flux, and

$$\dot{\varepsilon} = \frac{\partial \varepsilon}{\partial t} + \mathbf{v}.\nabla\varepsilon\,.$$

For a linear viscous fluid we take

$$\sigma_{ij} = -p\delta_{ij} + 2\mu d_{ij},$$

as in (2.3). In a general situation, the fluid will be compressible, but for many convective motions the system may be considerably simplified by assuming the motion is isochoric, i.e., essentially incompressible flow, except in the body force term $\mathbf{f}$ in $(2.5)_1$. This approximation has caused a lot of concern in the literature and is known as the *Boussinesq approximation*, associated with the work of Boussinesq (1897).

Analytical treatments of the Boussinesq approximation and a presentation of the conditions under which it is likely to be valid have been given by Spiegel and Veronis (1960), Mihaljan (1962), Fife (1970, 2000), (Chandrasekhar, 1981, pages 16–18), Breugem and Rees (2006), Barletta (2022), Barletta et al. (2023).

Our goal is to reduce the energy balance equation (2.8) to a form suitable for mathematical analysis. Hence, we now introduce the second law of thermodynamics in the form of the Clausius–Duhem inequality, namely,

$$\rho\dot{\eta} \geq \frac{\rho r}{T} - \left(\frac{q_i}{T}\right)_{,i} \tag{2.9}$$

where η is the entropy. We additionally introduce the Helmholtz free energy, ψ, which is defined by $\psi = \varepsilon - \eta T$. Now use equation (2.8) to remove the heat supply term r from inequality (2.9) to find

$$-\rho\dot{\psi} - \rho\eta\dot{T} + \sigma_{ij}d_{ij} - \frac{q_i T_{,i}}{T} \geq 0. \tag{2.10}$$

Since we are dealing with a viscous, incompressible fluid the dependent variables should be functions of the temperature, temperature gradient, and the symmetric part of the velocity gradient, and so the constitutive theory we utilize for the Helmholtz free energy function is that $\psi = \psi(T, T_{,i}, d_{ij})$. Upon employing the chain rule in inequality (2.10) one obtains

$$-\rho\left(\frac{\partial\psi}{\partial T} + \eta\right)\dot{T} - \rho\frac{\partial\psi}{\partial T_{,i}}\frac{d}{dt}(T_{,i}) - \frac{\partial\psi}{\partial d_{ij}}\frac{d}{dt}(d_{ij}) + \sigma_{ij}d_{ij} - \frac{q_i T_{,i}}{T} \geq 0, \tag{2.11}$$

where d/dt or a superposed dot denotes the material derivative. We may now argue that since $dT_{,i}/dt$ and $\dot{T}$ appear linearly in inequality (2.11) their coefficients must be zero, since by suitable choice of r the energy balance equation (2.8) may hold for an independent choice of $dT_{,i}/dt$ and $\dot{T}$. Thus, we must have

$$\psi = \psi(T, d_{ij}) \qquad \text{and} \qquad \eta = -\frac{\partial\psi}{\partial T}. \tag{2.12}$$

The fluid is incompressible and so $d(d_{ij})/dt$ is not independent for all six components. Thus we must introduce the constraint $d_{ii} = 0$ into inequality (2.11) via a Lagrange multiplier λ. Hence what remains of (2.11) may be written as

$$\frac{d}{dt}(d_{ij})\left(\lambda\delta_{ij} - \rho\frac{\partial\psi}{\partial d_{ij}}\right) + \sigma_{ij}d_{ij} - \frac{q_i T_{,i}}{T} \geq 0. \tag{2.13}$$

Now we may argue that all six components of $d(d_{ij})/dt$ are independent because λ is arbitrary, the momentum equation being balanced by a suitable body force f_i. Thus, since $d(d_{ij})/dt$ appears linearly in inequality (2.13),

$$\rho\frac{\partial\psi}{\partial d_{ij}} = \lambda\delta_{ij}. \tag{2.14}$$

Next, ψ is a scalar quantity and so it can only depend on scalar quantities derived from d_{ij}. Thus, by the Cayley–Hamilton theorem ψ will depend on the invariants of d_{ij}, i.e. on $I_1 = d_{ii}$, $I_2 = \text{tr}\,\mathbf{d}^2$, $I_3 = D$, where $D = \det \mathbf{d}$. Hence,

$$\psi = \psi(I_1, I_2, I_3).$$

Employ the relation $6D = \epsilon_{ijk}\epsilon_{abc}d_{ia}d_{jb}d_{kc}$ and then one finds $\partial I_3/\partial d_{ij} = f_{ij}$ where f_{ij} is a quadratic function in d_{ij}. Hence,

$$\frac{\partial \psi}{\partial d_{ij}} = \frac{\partial \psi}{\partial I_1}\delta_{ij} + 2\frac{\partial \psi}{\partial I_2}d_{ij} + \frac{\partial \psi}{\partial I_3}f_{ij}\,. \tag{2.15}$$

However, employing (2.14) we must have $\partial\psi/\partial I_2 = 0$, $\partial\psi/\partial I_3 = 0$. Then from (2.14) and (2.15)

$$\frac{\partial \psi}{\partial d_{ii}} = \frac{3\lambda}{\rho} = \frac{\partial \psi}{\partial I_1}$$

and so

$$\psi = \frac{3\lambda}{\rho}I_1 + \psi_0(T).$$

Thus we deduce $\psi = \psi(T)$. Since we know $\eta = -\partial\psi/\partial T$ it follows that $\eta = \eta(T)$. What remains from inequality (2.13) is

$$\sigma_{ij}d_{ij} - \frac{q_i T_{,i}}{T} \geq 0. \tag{2.16}$$

For a linearly viscous fluid obeying Fourier's law we know that $\sigma_{ij} = -p\delta_{ij} + 2\mu d_{ij}$ and $q_i = -kT_{,i}$. These expressions inserted into inequality (2.16) lead to

$$2\mu d_{ij}d_{ij} + k\frac{T_{,i}T_{,i}}{T} \geq 0\,.$$

Hence, necessarily we obtain the restrictions on the dynamic viscosity and on the thermal conductivity, entirely consistent with the physics, $\mu \geq 0$, $k \geq 0$.

To reduce the energy balance equation (2.8) we note $\varepsilon = \psi + \eta T$ so $\varepsilon = \varepsilon(T)$ and then we obtain

$$\rho T\dot{\eta} = \sigma_{ij}d_{ij} + \rho r - q_{i,i}\,. \tag{2.17}$$

Take now $r = 0$ and put $\eta = -\partial\psi/\partial T$ and employ the constitutive theory for σ_{ij} and q_i to obtain

$$-\rho T\psi_{TT}\dot{T} = (kT_{,i})_{,i} + 2\mu d_{ij}d_{ij}\,. \tag{2.18}$$

From thermodynamics we know

$$c_p = \frac{\partial \varepsilon}{\partial T}\bigg|_p + p\frac{\partial V}{\partial T}\bigg|_p$$

where the derivatives are at constant pressure, V denotes the specific volume, and c_p is the specific heat at constant pressure. Thus, in this case equation (2.18) may be written

$$\rho c_p \dot{T} = (kT_{,i})_{,i} + 2\mu d_{ij} d_{ij} \,. \tag{2.19}$$

It is usually argued that when one is interested in convection problems before the instability onset or not too far from the onset that one may discard the viscous dissipation term $2\mu d_{ij} d_{ij}$. Thus, for k constant we derive the reduced energy balance equation

$$\rho c_p \dot{T} = k\Delta T \,. \tag{2.20}$$

Of course, this supposes the velocity gradients are small. When one is studying strongly convective motion or certainly thermal turbulence, the $d_{ij} d_{ij}$ term should be retained.

For thermal convection with a linearly viscous incompressible fluid we thus have equation (2.20) governing the temperature field and equation $(2.5)_2$ expressing the solenoidal character of the velocity field. It remains to develop the momentum equation $(2.5)_1$ and this equation we rewrite as

$$\rho(v_{i,t} + v_j v_{i,j}) = -p_{,i} + \mu\Delta v_i + \rho f_i \,. \tag{2.21}$$

The pressure in (2.21) is regarded as an unknown to be solved for in the problem, and the density is assumed constant except in the body force term $\mathbf{f}$, where

$$\rho\mathbf{f} = -g\mathbf{k}\rho(T), \tag{2.22}$$

assuming the only force acting is gravity, and where $\mathbf{k} = (0, 0, 1)$. The density in (2.22) is expanded in a Taylor series, and at this stage we consider only the first term, so

$$\rho = \rho_0 \Big[1 - \alpha(T - T_R) \Big], \tag{2.23}$$

where ρ_0 is the density at temperature T_R, and α is the coefficient of thermal expansion given by

$$\alpha = -\frac{1}{\rho}\left(\frac{\partial\rho}{\partial T}\right)_{T_R} \,. \tag{2.24}$$

From this point we follow the approach in Fife (2000, pp. 72–74). Suppose now the fluid is contained in the layer $(x, y) \in \mathbb{R}^2 \times \{z \in (0, d)\}$ and

$$\big|\alpha(T - T_R)\big| << 1.$$

(This is not a serious restriction since the temperature field is normally restricted to a small interval and the coefficient of thermal expansion is usually very small.) Write equation (2.23) as $\rho = \rho_0 + \rho_1$ so that

$$\left|\frac{\rho_1}{\rho_0}\right| = \big|\alpha(T - T_R)\big| << 1.$$

Non-dimensionalize equation (2.21) with the time scale $\mathcal{T} = d^2\rho_0/\mu$, velocity scale $U = \mu/d\rho_0$, pressure scale $P = \mu^2/\rho_0 d^2$, and with $\mathbf{x} = \mathbf{x}^* d$, $p = Pp^*$, $T = T^\sharp T^*$. Drop the stars in what follows although it is now understood that the variables v_i, p, T

and x_i are non-dimensional. Then equation (2.21) may be rewritten as

$$\Big(1 + \frac{\rho_1}{\rho_0}\Big)(v_{i,t} + v_j v_{i,j}) = -p_{,i} + \Delta v_i - \frac{gd^3\rho_0^2}{\mu^2}\, k_i\big[1 - \alpha T^\sharp(T - T_R)\big], \tag{2.25}$$

where T_R is the non-dimensional form for T_R. Since $|\rho_1/\rho_0| << 1$ we discard the $(\rho_1/\rho_0)\dot{v}_i$ term. We cannot similarly discard the term in $g\alpha T^\sharp(T - T_R)$ since this is the driving force for the convective motion. Instead we argue as Fife (2000), p. 74, that we retain the matching orders of magnitude of the pressure and buoyancy terms. This means that we balance the buoyancy term involving $g\alpha T^\sharp$ by the pressure gradient.

For fixed upper and lower constant temperatures T_U and T_L we have in non-dimensional form a steady solution (where $\mathbf{v} \equiv 0$) for the temperature field of form $\bar{T} = -(T_L - T_U)z + T_L$. This leads to an ordinary differential equation for the steady pressure field $\bar{p}$,

$$\frac{d\bar{p}}{dz} = -\frac{gd^3\rho_0^2}{\mu^2}\left(1 - \alpha T^\sharp\big[-(T_L - T_U)z + T_L - T_R\big]\right).$$

Perturb the steady solution and now write $p = \bar{p} + \beta_1\pi$ and $T = \bar{T} + \gamma\theta$, where β_1 and γ are parameters in an expansion for p and T. Then from equation (2.25) for v_i a velocity perturbation we see that

$$v_{i,t} + v_j v_{i,j} = -\beta_1 \pi_{,i} + \frac{gd^3\rho_0^2\alpha\gamma}{\mu^2}\, k_i\theta + \Delta v_i\,.$$

We pick β_1 to balance the $\pi_{,i}$ and θ terms, namely, $\beta_1 = gd^3\rho_0^2\alpha\gamma/\mu^2$. Then pick γ such that $\gamma = \mu^2/gd^3\rho_0^2\alpha$. This leads to the equation

$$v_{i,t} + v_j v_{i,j} = -\pi_{,i} + k_i\theta + \Delta v_i\,. \tag{2.26}$$

This is the momentum equation after the Boussinesq approximation has been applied.

Thus, taking account of (2.20), (2.26), and $(2.5)_2$, the equations for a linearly viscous, heat conducting, incompressible fluid (utilizing the Boussinesq approximation) may be written as

$$\begin{aligned} v_{i,t} + v_j v_{i,j} &= -\frac{1}{\rho_0}\, p_{,i} + \nu\Delta v_i - k_i g\big(1 - \alpha[T - T_R]\big),\\ v_{i,i} &= 0,\\ T_{,t} + v_i T_{,i} &= \kappa\Delta T, \end{aligned} \tag{2.27}$$

where $\kappa = k/\rho_0 c_p$ is the thermal diffusivity.

In the above treatment we have taken the viscosity to be constant. For water between 0°C and 100°C (see e.g., Batchelor (1967) pp. 596, 597) c_p varies from 4.27 joule gm^{-1} deg C^{-1} to 4.216 joule gm^{-1} deg C^{-1}, κ varies from 1.33×10^{-3} cm^2 sec^{-1} to 1.66×10^{-3} cm^2 sec^{-1}, so it is certainly reasonable to treat them as constant in many convection studies. The kinematic viscosity ν varies from $1.787 \times 10^{-2}\,\mathrm{cm}^2\,\mathrm{sec}^{-1}$ to $0.295 \times 10^{-2}\,\mathrm{cm}^2\,\mathrm{sec}^{-1}$ so its variation is greater. We shall investigate thermal convection with ν constant, although nonlinear energy stability studies with a temperature dependent viscosity are discussed in Straughan (2023a), where several other pertinent references are given.

To be absolutely clear about the notation we rewrite equations (2.27) in terms of the velocity components, pressure and temperature u, v, w, p, T as

$$\begin{aligned}
&\frac{\partial u}{\partial t} + u\frac{\partial u}{\partial x} + v\frac{\partial u}{\partial y} + w\frac{\partial u}{\partial z} = -\frac{1}{\rho_0}\frac{\partial p}{\partial x} + \nu\Delta u,\\
&\frac{\partial v}{\partial t} + u\frac{\partial v}{\partial x} + v\frac{\partial v}{\partial y} + w\frac{\partial v}{\partial z} = -\frac{1}{\rho_0}\frac{\partial p}{\partial y} + \nu\Delta v,\\
&\frac{\partial w}{\partial t} + u\frac{\partial w}{\partial x} + v\frac{\partial w}{\partial y} + w\frac{\partial w}{\partial z} = -\frac{1}{\rho_0}\frac{\partial p}{\partial z} + \nu\Delta w + g\alpha T,\\
&\frac{\partial u}{\partial x} + \frac{\partial v}{\partial y} + \frac{\partial w}{\partial z} = 0,\\
&\frac{\partial T}{\partial t} + u\frac{\partial T}{\partial x} + v\frac{\partial T}{\partial y} + w\frac{\partial T}{\partial z} = \kappa\Delta T,
\end{aligned} \tag{2.28}$$

where the pressure in (2.27) has been replaced (without loss of generality) by the function $p = p + g(1 + \alpha T_R)z/\rho_0$.

2.4 Linear instability

Let us consider a linear viscous fluid governed by equations (2.27). The fluid occupies the domain contained between the horizontal planes $z = 0$ and $z = d$ with $(x, y) \in \mathbb{R}^2$. The upper plane is held at a fixed constant temperature T_U, while the lower plane is held at a fixed constant temperature T_L, with $T_L > T_U$.

In the above situation equations (2.27) possess the motionless conduction solution

$$\bar{v}_i \equiv 0, \qquad \bar{T} = \beta z + T_L, \tag{2.29}$$

where $\beta = (T_L - T_U)/d > 0$ and $\bar{p}(z)$ is determined up to a constant from equation $(2.27)_1$. To study the stability of solution (2.29) we introduce perturbations u_i, θ, π by $v_i = \bar{v}_i + u_i$, $T = \bar{T} + \theta$, $p = \bar{p} + \pi$ and we non-dimensionalize the resulting perturbation equations with the scalings of pressure $P = \rho_0\nu U/d$, time $\mathcal{T} = d^2/\nu$, length d, and velocity $U = \nu/d$. Introduce the Prandtl number $Pr = \nu/\kappa$ and the parameter $R = \sqrt{\alpha g \beta d^4/\kappa\nu}$ where $Ra = R^2$ is the Rayleigh number. The non-dimensional fully nonlinear perturbation equations may then be shown to be

$$\begin{aligned}
&u_{i,t} + u_j u_{i,j} = -\pi_{,i} + \Delta u_i + R\theta k_i,\\
&u_{i,i} = 0,\\
&Pr(\theta_{,t} + u_i\theta_{,i}) = Rw + \Delta\theta,
\end{aligned} \tag{2.30}$$

for $(x, y) \in \mathbb{R}^2$ and $z \in (0, 1)$, and where $w = u_3$. The boundary conditions to be satisfied are

$$u_i = 0, \qquad \theta = 0, \qquad \text{on } z = 0, 1, \tag{2.31}$$

together with u_i, π, θ satisfying a plane tiling periodicity in the (x, y) plane.

To analyse the linear instability of solution (2.29) we linearize (2.30) and then write

$$u_i = u_i(\mathbf{x})e^{\sigma t}, \qquad \pi = \pi(\mathbf{x})e^{\sigma t}, \qquad \theta = \theta(\mathbf{x})e^{\sigma t}.$$

The linear instability equations arising from (2.30) are

$$\begin{aligned} \sigma u_i &= -\pi_{,i} + \Delta u_i + R\theta k_i, \\ u_{i,i} &= 0, \\ \sigma Pr\theta &= Rw + \Delta\theta. \end{aligned} \tag{2.32}$$

Let V be a periodic cell for the solution. Let now $\|\cdot\|$ and $(\cdot, \cdot)$ denote the norm and inner product on $L^2(V)$ where the variables may be complex.

2.4.1 Exchange of stabilities

Multiply $(2.32)_1$ by u_i^*, and integrate over V where $*$ denotes the complex conjugate. Likewise, multiply $(2.32)_3$ by θ^*, and integrate over V. One derives

$$\begin{aligned} \sigma\|\mathbf{u}\|^2 &= -\|\nabla\mathbf{u}\|^2 + R(\theta, w^*), \\ \sigma Pr\|\theta\|^2 &= R(w, \theta^*) - \|\nabla\theta\|^2, \end{aligned}$$

where we have integrated by parts and employed the boundary conditions. These equations are added to find

$$\sigma(\|\mathbf{u}\|^2 + \|\theta\|^2) = -\|\nabla\mathbf{u}\|^2 - \|\nabla\theta\|^2 + R[(\theta, w^*) + (w, \theta^*)]. \tag{2.33}$$

The right hand side of this equation is real. Upon settting $\sigma = \sigma_r + i\sigma_1$ and taking the imaginary part of (2.33) one sees that

$$\sigma_1(\|\mathbf{u}\|^2 + \|\theta\|^2) = 0. \tag{2.34}$$

Since we do not want a zero solution it follows that $\sigma_1 = 0$, and so the principle of exchange of stabilities holds. This is important, since σ is the growth rate of the solution to the linearized problem it means that to find the threshold for linear instability one may put $\sigma = 0$ in (2.32) and then solve the result as an eigenvalue problem for R.

We have explicitly included this demonstration of exchange of stabilities because if it does not hold then at the boundary where linear instability commences one knows $\sigma_r = 0$, but it may be that $\sigma_1 \neq 0$ and this means instability commences as oscillatory convection, or as a Hopf bifurcation. This is especially important in the current book since we shall see that when Fourier's law is replaced by a Maxwell–Cattaneo one then oscillatory convection is possible for a certain range of parameters. The addition of the $\tau q_{i,t}$ term in the Maxwell–Cattaneo relation leads to a second order in time equation for the temperature field and this in turn may lead to oscillatory solutions, cf. oscillatory solutions to the wave equation.

To solve (2.32) with $\sigma = 0$ we take curlcurl of $(2.32)_1$ and retain the resulting third component. This yields the system of equations

$$\begin{aligned} \Delta^2 w + R\Delta^*\theta &= 0, \\ \Delta\theta + Rw &= 0, \end{aligned} \tag{2.35}$$

where $\Delta^* = \partial^2/\partial x^2 + \partial^2\partial y^2$ is the horizontal Laplacian. This system is to be solved with the boundary conditions

$$w = 0, \quad \theta = 0, \qquad z = 0, 1,$$

recalling w, θ must also satisfy a plane tiling periodicity in the x, y plane. However, two further boundary conditions are required and these are, cf. Chandrasekhar (1981),

$$\frac{\partial w}{\partial z} = 0, \qquad z = 0, 1,$$

for two fixed surfaces, or

$$\frac{\partial^2 w}{\partial z^2} = 0, \qquad z = 0, 1,$$

for two stress free surfaces.

In general, for two fixed surfaces it is preferable to obtain a numerical solution for the eigenvalues R. One may obtain an analytical solution for the current problem and details are given in (Chandrasekhar, 1981, pages 36–43).

The two stress free surfaces case is often employed to illustrate solution behaviour and this case is handled by a normal mode method, as explained clearly by Chandrasekhar (1981), and by Barletta (2019, 2021).

Since we only need one solution to be unstable we write

$$w = W(z)h(x, y), \qquad \theta = \Theta(z)h(x, y),$$

where h is a function which satisfies the equation $\Delta^* h = -a^2 h$, a being a wavenumber. The specific forms h may have are triangles, rectangles, rolls, or hexagons and specific forms are presented in very clear detail in (Chandrasekhar, 1981, pages 43–52). With the above representation (2.35) become

$$\begin{aligned} (D^2 - a^2)^2 W - Ra^2\Theta &= 0, \\ (D^2 - a^2)\Theta + RW &= 0, \end{aligned} \tag{2.36}$$

where $D = d/dz$. These equations are to be solved subject to the boundary conditions

$$W = 0, \quad \Theta = 0, \quad DW = 0, \qquad z = 0, 1, \tag{2.37}$$

for two fixed surfaces, or

$$W = 0, \quad \Theta = 0, \quad D^2W = 0, \qquad z = 0, 1, \tag{2.38}$$

for two stress free surfaces.

A very convenient way to solve (2.36) with (2.37) is by a Chebyshev tau numerical method, see e.g. Dongarra et al. (1996).

For two stress free surfaces we may write

$$W = \sum_{n=1}^{\infty} W_n \sin(n\pi z), \qquad \Theta = \sum_{n=1}^{\infty} \Theta_n \sin(n\pi z),$$

and then (2.36) reduce to the determinant equation

$$\begin{vmatrix} \Lambda_n^2 & -Ra^2 \\ R & -\Lambda_n \end{vmatrix} = 0$$

where $\Lambda_n = n^2\pi^2 + a^2$. Thus $R^2 = \Lambda_n^3/a^2$. One requires the minimum of $R^2(n^2, a^2)$, and by taking $\partial R^2/\partial n^2$ it follows $n = 1$ achieves this. Then $R^2 = \Lambda^3/a^2$, where $\Lambda = \pi^2 + a^2$. Minimization of this in a^2 yields the critical values

$$a_{crit}^2 = \frac{\pi^2}{2}, \qquad R_{crit}^2 = Ra_{crit} = \frac{27\pi^4}{4},$$

as shown by Chandrasekhar (1981).

2.5 Global nonlinear stability

To analyse nonlinear stability we let $\|\cdot\|$ and $(\cdot,\cdot)$ be the norm and inner product on $L^2(V)$.

Multiply $(2.30)_1$ by u_i and integrate over V. Multiply $(2.30)_3$ by θ and integrate over V. After some integration by parts and use of the boundary conditions (2.31) one may derive

$$\begin{aligned} \frac{d}{dt}\frac{1}{2}\|\mathbf{u}\|^2 &= -\|\nabla\mathbf{u}\|^2 + R(\theta, w), \\ \frac{d}{dt}\frac{1}{2}\|\theta\|^2 &= -\|\nabla\theta\|^2 + R(\theta, w). \end{aligned} \tag{2.39}$$

Define the energy function, $E(t)$, the production term, $I(t)$, and the dissipation function, $D(t)$, by

$$\begin{aligned} E &= \frac{1}{2}\|\mathbf{u}\|^2 + \frac{Pr}{2}\|\theta\|^2, \\ I &= 2(\theta, w), \\ D &= \|\nabla\mathbf{u}\|^2 + \|\nabla\theta\|^2. \end{aligned} \tag{2.40}$$

Then, upon adding the equations in (2.39) one arrives at an energy equation

$$\frac{dE}{dt} = RI - D. \tag{2.41}$$

Define R_E by

$$\frac{1}{R_E} = \max_H \frac{I}{D}, \tag{2.42}$$

where H is the space of admissible solutions,

$$H = \{\mathbf{u}, \theta | u_i \in H^1(V), \theta \in H^1(V), u_{i,i} = 0\},$$

where H includes the relevant boundary conditions. Observe that from (2.41)

$$\frac{dE}{dt} \leq -D\left(1 - \frac{R}{R_E}\right). \tag{2.43}$$

Suppose now $b = R/R_E < 1$, and then using Poincaré's inequality one may show $D \geq 2k\pi^2 E$, where $k = \min\{1, Pr^{-1}\}$, and hence from (2.43) one deduces

$$\frac{dE}{dt} \leq -2k\pi^2 E. \tag{2.44}$$

Integration of this shows

$$E(t) \leq E(0)\exp(-2k\pi^2 t). \tag{2.45}$$

Hence, provided $R < R_E$, $E(t)$ decays rapidly and global nonlinear stability holds. Global in this case means for all initial data.

It remains to find R_E and the Euler - Lagrange equations from (2.42) are

$$\begin{aligned} \Delta u_i + R_E k_i \theta &= \lambda_{,i}, \\ u_{i,i} &= 0, \\ \Delta\theta + R_E w &= 0, \end{aligned} \tag{2.46}$$

where $\lambda(\mathbf{x})$ is a Lagrange multiplier. Let us observe that (2.46) are exactly the same as the linear instability equations (2.32) once exchange of stabilities is applied. Hence, the eigenvalue problem governing nonlinear stability is identical to the one controlling linear instability.

Therefore, it follows that if $Ra < Ra_{crit}$ then global nonlinear stability holds, whereas if $Ra > Ra_{crit}$ there is linear instability, where $Ra = R^2$. In this case the critical Rayleigh numbers of linear instability and nonlinear stability, Ra_L and Ra_E, respectively, are the same.

We include this section to illustrate the energy method. However, we point out that the result seen here is special in that coincidence of the linear and nonlinear stability boundaries does not normally occur. Once a Cattaneo like theory is employed oscillatory instabilities are found and then we normally find that $Ra_E < Ra_L$. Of course, if they are relatively close this is still very useful.

2.6 Navier–Stokes–Voigt theory

This section is included because with some of the Cattaneo models it appears necessary to have extra regularizing terms present to control the nonlinear contributions which arise. A Navier–Stokes–Voigt theory is capable of achieving this.

We believe the Navier–Stokes–Voigt theory of thermal convection is of interest in its own right and merits inclusion here. Recent work (which has arisen from the Russian literature) has focussed attention on a particular class of viscoelastic fluids known as Kelvin–Voigt fluids. Actually, there are Kelvin–Voigt fluids or order N, $N = 0, 1, \ldots$ The Kelvin–Voigt fluid of order 0 is also known as a Navier–Stokes–Voigt fluid, and this has been analysed extensively in recent work, see e.g. Kalantarov and Titi (2009), Kalantarov et al. (2009), Berselli and Bisconti (2012), Layton and Rebholz (2013), Damázio et al. (2016), Straughan (2021b). General classes of Kelvin–Voigt fluids have been investigated by Oskolkov (1988, 1995), Oskolkov and Shadiev (1994), and thermal effects are included in the analyses by Sukacheva and Matveeva (2010), Matveeva (2013), Sukacheva and Kondyukov (2014). Particular classes of thermal convection problems in Kelvin–Voigt fluids of order N are investigated by Straughan (2021a, 2022b). The models of Kelvin–Voigt, Maxwell, Oldroyd, and Stokes fluids are compared in Straughan (2022a), where the issue raised by Damázio et al. (2016), on the significance of the Kelvin–Voigt term in a Navier–Stokes–Voigt theory is also discussed.

For the purpose of this book, a Navier–Stokes–Voigt fluid in a thermal convection setting adds a regularizing term involving the Laplacian of the velocity field into the momentum equation, and the appropriate equations are, cf. the classical equations (2.27),

$$\begin{aligned} v_{i,t} + v_j v_{i,j} - \lambda \Delta v_{i,t} &= -\frac{1}{\rho_0} p_{,i} + \nu \Delta v_i - k_i g\big(1 - \alpha[T - T_R]\big), \\ v_{i,i} &= 0, \\ T_{,t} + v_i T_{,i} &= \kappa \Delta T, \end{aligned} \tag{2.47}$$

where $\lambda > 0$ is the Kelvin–Voigt coefficient.

The steady state solution is again (2.29). Hence, the noninear perturbation equations for a stability analysis are

$$\begin{aligned} &u_{i,t} + u_j u_{i,j} - \lambda \Delta u_{i,t} = -\pi_{,i} + \Delta u_i + R\theta k_i, \\ &u_{i,i} = 0, \\ &Pr(\theta_{,t} + u_i \theta_{,i}) = Rw + \Delta\theta, \end{aligned} \tag{2.48}$$

together with the boundary conditions (2.31) and the periodicity requirements following (2.31).

For system (2.48) one may again show exchange of stabilities holds. Instead of equation (2.34) one now finds

$$\sigma_1(\|\mathbf{u}\|^2 + \lambda\|\nabla\mathbf{u}\|^2 + \|\theta\|^2) = 0,$$

and the outcome is the same, i.e. $\sigma \in \mathbb{R}$. Thus, the linear instability analysis follows exactly that in section 2.4, leading to the same critical Rayleigh and wave numbers as found there.

The nonlinear stability analysis proceeds as in section 2.5 except the energy function is now

$$E = \frac{1}{2}\|\mathbf{u}\|^2 + \frac{\lambda}{2}\|\nabla\mathbf{u}\|^2 + \frac{Pr}{2}\|\theta\|^2.$$

Inequality (2.43) continues to hold, but now

$$\begin{aligned} D &= \|\nabla\mathbf{u}\|^2 + \|\nabla\theta\|^2 \\ &\geq \frac{1}{2}\|\nabla\mathbf{u}\|^2 + \frac{\pi^2}{2}\|\mathbf{u}\|^2 + \pi^2\|\theta\|^2 \\ &\geq k_1 E, \end{aligned}$$

where $k_1 = \min\{\lambda^{-1}, \pi^2, 2\pi^2/Pr\}$. One now obtains inequality (2.44) with k_1 instead of $2k\pi^2$, and again global nonlinear stability follows if $R < R_E$. Since I/D is the same as in section 2.5 it follows that the linear and nonlinear stability Rayleigh number thresholds are the same.

Chapter 3
Thermal convection with a Cattaneo flux law

3.1 Thermal convection models

In this chapter the aim is to extend the basic equations for thermal convection in a linearly viscous incompressible fluid, namely equations (2.27), which are based on Fourier's law, $q_i = -k\,\partial T/\partial x_i$, to incorporate the Cattaneo flux law $(1.40)_2$. For the fluid case the particles can move and so equation $(1.40)_2$ must be extended to allow for the fluid velocity. The partial time derivative of the heat flux, $\partial q/\partial t$, must be replaced by a suitable derivative incorporating fluid movement. Since $q_i = q_i(\mathbf{x}, t)$ and $\mathbf{x} = \mathbf{x}(\mathbf{X}, t)$, where $\mathbf{X}$ refers to particles in a reference configuration, a natural derivative to employ is the material derivative

$$\dot{q}_i \equiv \frac{Dq_i}{Dt} = \frac{\partial q_i}{\partial t} + v_j \frac{\partial q_i}{\partial x_j} .$$

Thus, $(1.40)_2$ would be replaced by the equation

$$\tau \frac{Dq_i}{Dt} + q_i = -k \frac{\partial T}{\partial x_i} . \tag{3.1}$$

Such an approach is advocated by Christov and Jordan (2005) and is consistent with the premise of extended thermodynamics where this equation is regarded as a balance law which governs the behaviour of q_i, see Ruggeri (1983), Jou et al. (2010), Ruggeri and Sugiyama (2021). Within thermal convection such a derivative was employed by Bissell (2015).

To employ (3.1) is one approach to the problem. Another avenue is to argue that Fourier's law is a constitutive equation and since (3.1) replaces this then the replacement should also be a constitutive equation. However, in this case one may argue that since the material derivative is not an objective quantity one should replace (3.1) by an equation of form

$$\tau \frac{\mathcal{D}q_i}{\mathcal{D}t} + q_i = -k \frac{\partial T}{\partial x_i} , \tag{3.2}$$

B. Straughan, *Thermal Convection with a Cattaneo Flux Law*, Advances in Mechanics and Mathematics 53, https://doi.org/10.1007/978-3-031-80885-2_3

where $\mathcal{D}q_i/\mathcal{D}t$ is a suitable objective derivative. Objective derivatives in this context are discussed at length by Capriz and Mariano (2014), Morro (2017a,b, 2018, 2022), and by Morro and Giorgi (2020).

Gentile and Straughan (2024) investigate the problem of thermal convection in a linearly viscous incompressible fluid using (3.1) or (3.2), and we follow and elaborate on their work here.

One of the early models of fluid dynamics which incorporated a Cattaneo theory for the heat flux was that of Fox (1969). He employed what Morro (2022) refers to as a co-rotational derivative and the theory of Fox (1969) relies on the objective derivative

$$\frac{\mathcal{D}Q_i}{\mathcal{D}t} = \frac{\partial Q_i}{\partial t} + v_j\frac{\partial Q_i}{\partial x_j} + \frac{1}{2}\Big(Q_j\frac{\partial v_j}{\partial x_i} - Q_j\frac{\partial v_i}{\partial x_j}\Big). \tag{3.3}$$

We at this point use Q_i to denote the heat flux since in our subsequent stability analyses lower case q_i is employed to denote a perturbation to a steady state heat flux. The theory of Fox (1969) was employed in what we believe was the first analysis of thermal convection with a Cattaneo effect by Straughan and Franchi (1984).

Another objective derivative which Morro (2022) refers to as the Truesdell derivative, Truesdell (1955), or the Oldroyd derivative, is

$$\frac{\mathcal{D}Q_i}{\mathcal{D}t} = \frac{\partial Q_i}{\partial t} + v_j\frac{\partial Q_i}{\partial x_j} - Q_j\frac{\partial v_i}{\partial x_j}. \tag{3.4}$$

This derivative was independently suggested by Christov (2009) and it was employed by Straughan (2010b) in an analysis of thermal convection. The resulting theory is now usually referred to as Cattaneo–Christov convection.

A further objective derivative to consider is the Cotter–Rivlin one, see Morro (2022), and this is

$$\frac{\mathcal{D}Q_i}{\mathcal{D}t} = \frac{\partial Q_i}{\partial t} + v_j\frac{\partial Q_i}{\partial x_j} + Q_j\frac{\partial v_j}{\partial x_i}. \tag{3.5}$$

It is interesting that this derivative is extended to compressible fluid flow by Angeles (2023). He shows that *for the compressible case* where $v_{i,i} \neq 0$ and the objective derivative also contains this effect, the only derivatives which guarantee that the system is hyperbolic are the material derivative or the analogous Cotter–Rivlin one with an appropriate addition of the divergence of the velocity field. In this work, however, we only consider an incompressible fluid. The Cotter–Rivlin derivative in the context of thermal convection is considered in this chapter.

The general form of objective derivative in the incompressible case is given by Morro (2018, 2022) as

$$\begin{aligned}\frac{\mathcal{D}Q_i}{\mathcal{D}t} &= \frac{\partial Q_i}{\partial t} + v_j\frac{\partial Q_i}{\partial x_j} - \frac{Q_j}{2}\Big(\frac{\partial v_i}{\partial x_j} - \frac{\partial v_j}{\partial x_i}\Big) + \frac{\gamma Q_j}{2}\Big(\frac{\partial v_i}{\partial x_j} + \frac{\partial v_j}{\partial x_i}\Big),\\ &= \frac{\partial Q_i}{\partial t} + v_j\frac{\partial Q_i}{\partial x_j} - \frac{(1-\gamma)}{2}Q_j\frac{\partial v_i}{\partial x_j} + \frac{(1+\gamma)}{2}Q_j\frac{\partial v_j}{\partial x_i},\end{aligned} \tag{3.6}$$

where for our purpose γ is constant. Equation (3.6) may be conveniently written as

$$\frac{\mathcal{D}Q_i}{\mathcal{D}t} = \frac{\partial Q_i}{\partial t} + v_j \frac{\partial Q_i}{\partial x_j} - W_{ij}Q_j + \gamma D_{ij}Q_j\,, \tag{3.7}$$

where W_{ij} and D_{ij} are the skew - symmetric and symmetric parts of the velocity gradient, i.e.

$$W_{ij} = \frac{1}{2}\Big(\frac{\partial v_i}{\partial x_j} - \frac{\partial v_j}{\partial x_i}\Big), \qquad D_{ij} = \frac{1}{2}\Big(\frac{\partial v_i}{\partial x_j} + \frac{\partial v_j}{\partial x_i}\Big).$$

This objective derivative is analysed for thermal convection by Gentile and Straughan (2024).

The basic equations for thermal convection using the material derivative (3.1) are known as material derivative theory, or Cattaneo–Christov–Jordan (CCJ) theory, after Christov and Jordan (2005), see also Ruggeri (1983), Ruggeri and Sugiyama (2021). Those equations employing the co-rotational derivative (3.3) are referred to as Cattaneo–Fox theory. The equations utilizing (3.4) are referred to as Cattaneo–Christov theory. When (3.5) is used the resulting system is called Cattaneo–Cotter–Rivlin (CCR) theory. Finally, if the general objective derivative (3.6) is used the resulting theory is called Cattaneo–Morro theory.

The general form of equations to represent thermal convection when the heat flux law is of Cattaneo type, generalizing (2.27) are

$$\begin{aligned}
&\frac{\partial v_i}{\partial t} + v_j \frac{\partial v_i}{\partial x_j} = -\frac{1}{\rho_0}\frac{\partial p}{\partial x_i} + \nu\Delta v_i + \alpha g k_i T,\\
&\frac{\partial v_i}{\partial x_i} = 0,\\
&\frac{\partial T}{\partial t} + v_i \frac{\partial T}{\partial x_i} = -\frac{\partial Q_i}{\partial x_i},\\
&\tau\frac{\mathcal{D}Q_i}{\mathcal{D}t} + Q_i = -k\frac{\partial T}{\partial x_i},
\end{aligned} \tag{3.8}$$

where the terms are as in (2.27) and $\mathcal{D}Q_i/\mathcal{D}t$ is replaced by one of the derivatives of this section.

In the interests of clarity we give specific details of each system.

3.1.1 Material derivative, CCJ theory

When (3.1) is employed the equations of material derivative theory, or Cattaneo–Christov–Jordan theory are

$$\begin{aligned}
&\frac{\partial v_i}{\partial t} + v_j \frac{\partial v_i}{\partial x_j} = -\frac{1}{\rho_0} \frac{\partial p}{\partial x_i} + \nu \Delta v_i + \alpha g k_i T, \\
&\frac{\partial v_i}{\partial x_i} = 0, \\
&\frac{\partial T}{\partial t} + v_i \frac{\partial T}{\partial x_i} = -\frac{\partial Q_i}{\partial x_i}, \\
&\tau \Big(\frac{\partial Q_i}{\partial t} + v_j \frac{\partial Q_i}{\partial x_j} \Big) + Q_i = -k \frac{\partial T}{\partial x_i}\,.
\end{aligned} \tag{3.9}$$

3.1.2 Cattaneo–Fox theory

If one uses equation (3.3) then the system representing Cattaneo–Fox theory is

$$\begin{aligned}
&\frac{\partial v_i}{\partial t} + v_j \frac{\partial v_i}{\partial x_j} = -\frac{1}{\rho_0} \frac{\partial p}{\partial x_i} + \nu \Delta v_i + \alpha g k_i T, \\
&\frac{\partial v_i}{\partial x_i} = 0, \\
&\frac{\partial T}{\partial t} + v_i \frac{\partial T}{\partial x_i} = -\frac{\partial Q_i}{\partial x_i}, \\
&\tau \left(\frac{\partial Q_i}{\partial t} + v_j \frac{\partial Q_i}{\partial x_j} + \frac{Q_j}{2} \left[\frac{\partial v_j}{\partial x_i} - \frac{\partial v_i}{\partial x_j} \right] \right) + Q_i = -k \frac{\partial T}{\partial x_i}\,.
\end{aligned} \tag{3.10}$$

3.1.3 Cattaneo–Christov theory

When equation (3.4) is employed then the Cattaneo–Christov system is

$$\begin{aligned}
&\frac{\partial v_i}{\partial t} + v_j \frac{\partial v_i}{\partial x_j} = -\frac{1}{\rho_0} \frac{\partial p}{\partial x_i} + \nu \Delta v_i + \alpha g k_i T, \\
&\frac{\partial v_i}{\partial x_i} = 0, \\
&\frac{\partial T}{\partial t} + v_i \frac{\partial T}{\partial x_i} = -\frac{\partial Q_i}{\partial x_i}, \\
&\tau \Big(\frac{\partial Q_i}{\partial t} + v_j \frac{\partial Q_i}{\partial x_j} - Q_j \frac{\partial v_i}{\partial x_j} \Big) + Q_i = -k \frac{\partial T}{\partial x_i}\,.
\end{aligned} \tag{3.11}$$

3.1.4 CCR theory

The system of equations obtained using (3.5) is that of Cattaneo–Cotter–Rivlin theory, namely

$$
\begin{aligned}
&\frac{\partial v_i}{\partial t} + v_j \frac{\partial v_i}{\partial x_j} = -\frac{1}{\rho_0}\frac{\partial p}{\partial x_i} + \nu \Delta v_i + \alpha g k_i T,\\
&\frac{\partial v_i}{\partial x_i} = 0,\\
&\frac{\partial T}{\partial t} + v_i \frac{\partial T}{\partial x_i} = -\frac{\partial Q_i}{\partial x_i},\\
&\tau\Big(\frac{\partial Q_i}{\partial t} + v_j \frac{\partial Q_i}{\partial x_j} + Q_j \frac{\partial v_j}{\partial x_i}\Big) + Q_i = -k\frac{\partial T}{\partial x_i}.
\end{aligned}
\tag{3.12}
$$

3.1.5 Cattaneo–Morro theory

When the general objective derivative form (3.6) is utilized one obtains the Cattaneo–Morro system,

$$
\begin{aligned}
&\frac{\partial v_i}{\partial t} + v_j \frac{\partial v_i}{\partial x_j} = -\frac{1}{\rho_0}\frac{\partial p}{\partial x_i} + \nu \Delta v_i + \alpha g k_i T,\\
&\frac{\partial v_i}{\partial x_i} = 0,\\
&\frac{\partial T}{\partial t} + v_i \frac{\partial T}{\partial x_i} = -\frac{\partial Q_i}{\partial x_i},\\
&\tau\Big(\frac{\partial Q_i}{\partial t} + v_j \frac{\partial Q_i}{\partial x_j} - W_{ij}Q_j + \gamma D_{ij}Q_j\Big) + Q_i = -k\frac{\partial T}{\partial x_i}.
\end{aligned}
\tag{3.13}
$$

For the purpose of this work γ is any real number.

3.2 Steady, conduction only solution

The steady solution to all of the models (3.9) - (3.13) is essentially (2.29) although now the steady heat flux enters the calculation. Thus, the steady solution corresponding to the boundary conditions of section 2.4 is

$$
\bar{v}_i \equiv 0, \qquad \bar{T} = -\beta z + T_L\,, \qquad \bar{\mathbf{Q}} = (0, 0, k\beta). \tag{3.14}
$$

The perturbation quantities are introduced as in section 2.4 although now the perturbation to $\bar{Q}_i$ is q_i. The non-dimensional variables are exactly as in section 2.4 with the addition of the parameter Sg introduced by Papanicolaou et al. (2011), namely,

$$Sg = \frac{\tau \nu}{d^2}. \tag{3.15}$$

The non-dimensional perturbation boundary conditions are (2.31).

After some calculation the fully nonlinear, non-dimensional perturbation equations may be written for each system as below, cf. Gentile and Straughan (2024).

3.2.1 Perturbation equations, CCJ theory

For the CCJ theory the nonlinear perturbation equations are

$$\begin{aligned} &u_{i,t} + u_j u_{i,j} = -\pi_{,i} + \Delta u_i + R k_i \theta, \\ &u_{i,i} = 0, \\ &Pr(\theta_{,t} + u_i \theta_{,i}) = Rw - q_{i,i}, \\ &Sg(q_{i,t} + u_j q_{i,j}) = -\theta_{,i} - q_i, \end{aligned} \tag{3.16}$$

3.2.2 Perturbation equations, Cattaneo–Fox theory

For the Cattaneo–Fox theory the nonlinear perturbation equations are

$$\begin{aligned} &u_{i,t} + u_j u_{i,j} = -\pi_{,i} + \Delta u_i + R k_i \theta, \\ &u_{i,i} = 0, \\ &Pr(\theta_{,t} + u_i \theta_{,i}) = Rw - q_{i,i}, \\ &Sg\Big(q_{i,t} + u_j q_{i,j} - \frac{1}{2} u_{i,j} q_j + \frac{1}{2} u_{j,i} q_j\Big) = \\ &\qquad - \theta_{,i} - q_i + \frac{SgR}{2Pr} u_{i,3} - \frac{SgR}{2Pr} w_{,i}. \end{aligned} \tag{3.17}$$

3.2.3 Perturbation equations, Cattaneo–Christov theory

For the Cattaneo–Christov theory the nonlinear perturbation equations are

$$\begin{aligned} &u_{i,t} + u_j u_{i,j} = -\pi_{,i} + \Delta u_i + R k_i \theta, \\ &u_{i,i} = 0, \\ &Pr(\theta_{,t} + u_i \theta_{,i}) = Rw - q_{i,i}, \\ &Sg(q_{i,t} + u_j q_{i,j} - u_{i,j} q_j) = -\theta_{,i} - q_i + \frac{SgR}{Pr} u_{i,3}. \end{aligned} \tag{3.18}$$

3.2.4 Perturbation equations, CCR theory

For the CCR theory the nonlinear perturbation equations are

$$\begin{aligned} u_{i,t} + u_j u_{i,j} &= -\pi_{,i} + \Delta u_i + R k_i \theta, \\ u_{i,i} &= 0, \\ Pr(\theta_{,t} + u_i \theta_{,i}) &= Rw - q_{i,i}\,, \\ Sg(q_{i,t} + u_j q_{i,j} + u_{j,i} q_j) &= -\theta_{,i} - q_i - \frac{SgR}{Pr} w_{,i}. \end{aligned} \tag{3.19}$$

3.2.5 Perturbation equations, Cattaneo–Morro theory

For the Cattaneo–Morro theory the nonlinear perturbation equations are

$$\begin{aligned} & u_{i,t} + u_j u_{i,j} = -\pi_{,i} + \Delta u_i + R k_i \theta, \\ & u_{i,i} = 0, \\ & Pr(\theta_{,t} + u_i \theta_{,i}) = Rw - q_{i,i}\,, \\ & Sg\Big(q_{i,t} + u_j q_{i,j} - \Big[\frac{1-\gamma}{2}\Big] u_{i,j} q_j + \Big[\frac{1+\gamma}{2}\Big] u_{j,i} q_j\Big) = \\ & \qquad -\theta_{,i} - q_i + \frac{SgR}{2Pr}(1-\gamma) u_{i,3} - \frac{SgR}{2Pr}(1+\gamma) w_{,i}. \end{aligned} \tag{3.20}$$

All systems (3.16) - (3.20) hold on the domain $\mathbb{R}^2 \times \{z \in (0,1)\} \times \{t > 0\}$, and satisfy the boundary conditions

$$u_i = 0, \quad \theta = 0, \qquad z = 0, 1, \tag{3.21}$$

together with the fact that u_i, θ, π, q_i satisfy plane tiling periodic boundary conditions in x, y. The forms of the convection cell resulting from the periodicity are triangles, rectangles and hexagons, and full details may be found in Chandrasekhar (1981, pages 43–53).

3.3 Linear instability

Before we discuss the linear instability analyses of equations (3.16) - (3.20) we point out one very important difference between each of these sets of equations and the system (2.30) which governs classical thermal convection with Fourier theory.

If one ignores the nonlinear terms and the time derivatives, the remaining linear operator for (2.30) is a symmetric operator and then oscillatory instabilities do not arise, cf. Galdi and Straughan (1985a). However, for each of (3.16) - (3.20) the terms $-q_{i,i}$ and $-\theta_{,i}$ in the temperature and heat flux equations, respectively, define a skew - symmetric operator. For example, for (3.16) the linear operator acting on $(u, v, w, \theta, q_1, q_2, q_3)^T$ is seen to have form

$$L = \begin{pmatrix} \Delta & 0 & 0 & 0 & 0 & 0 & 0 \\ 0 & \Delta & 0 & 0 & 0 & 0 & 0 \\ 0 & 0 & \Delta & R & 0 & 0 & 0 \\ 0 & 0 & R & 0 & -\dfrac{\partial}{\partial x} & -\dfrac{\partial}{\partial y} & -\dfrac{\partial}{\partial z} \\ 0 & 0 & 0 & -\dfrac{\partial}{\partial x} & -1 & 0 & 0 \\ 0 & 0 & 0 & -\dfrac{\partial}{\partial y} & 0 & -1 & 0 \\ 0 & 0 & 0 & -\dfrac{\partial}{\partial z} & 0 & 0 & -1 \end{pmatrix}$$

Thus L may be written as

$$L = L_S + L_A ,$$

L_S being a symmetric operator, while L_A is skew-symmetric, where

$$L_S = \begin{pmatrix} \Delta & 0 & 0 & 0 & 0 & 0 & 0 \\ 0 & \Delta & 0 & 0 & 0 & 0 & 0 \\ 0 & 0 & \Delta & R & 0 & 0 & 0 \\ 0 & 0 & R & 0 & 0 & 0 & 0 \\ 0 & 0 & 0 & 0 & -1 & 0 & 0 \\ 0 & 0 & 0 & 0 & 0 & -1 & 0 \\ 0 & 0 & 0 & 0 & 0 & 0 & -1 \end{pmatrix}$$

and

$$L_A = \begin{pmatrix} 0 & 0 & 0 & 0 & 0 & 0 & 0 \\ 0 & 0 & 0 & 0 & 0 & 0 & 0 \\ 0 & 0 & 0 & 0 & 0 & 0 & 0 \\ 0 & 0 & 0 & 0 & -\dfrac{\partial}{\partial x} & -\dfrac{\partial}{\partial y} & -\dfrac{\partial}{\partial z} \\ 0 & 0 & 0 & -\dfrac{\partial}{\partial x} & 0 & 0 & 0 \\ 0 & 0 & 0 & -\dfrac{\partial}{\partial y} & 0 & 0 & 0 \\ 0 & 0 & 0 & -\dfrac{\partial}{\partial z} & 0 & 0 & 0 \end{pmatrix}$$

If $U^1 = (u^1, v^1, w^1, \theta^1, q_1^1, q_2^1, q_3^1)^T$ and $U^2 = (u^2, v^2, w^2, \theta^2, q_1^2, q_2^2, q_3^2)^T$ are two solutions satisfying the boundary conditions then for the inner product $(\cdot, \cdot)$ over a period cell V one has

$$(L_A U^1, U^2) = -\Big(\frac{\partial q_1^1}{\partial x}, \theta^2\Big) - \Big(\frac{\partial q_2^1}{\partial y}, \theta^2\Big) - \Big(\frac{\partial q_3^1}{\partial z}, \theta^2\Big)$$
$$- \Big(\frac{\partial \theta^1}{\partial x}, q_1^2\Big) - \Big(\frac{\partial \theta^1}{\partial y}, q_2^2\Big) - \Big(\frac{\partial \theta^1}{\partial z}, q_3^2\Big).$$

Integrations by parts and use of the boundary conditions and periodicity shows that

$$(L_A U^2, U^1) = -(L_A U^1, U^2),$$

confirming L_A is skew - symmetric. Hence, one cannot rule out oscillatory instabilities with the Cattaneo models, and, indeed, this is what we find.

This is important, because we do not expect to find a nonlinear energy stability analysis will yield exactly the same stability boundary as that found by a normal mode linear instability theory. The second order nature of the time derivatives of temperature are inducing a very different behaviour to that found in classical Bénard (Fourier) convection.

Whichever of the systems (3.16) - (3.20) are chosen for a linear instability analysis the appropriate form of equations are linearized and then one selects

$$u_i = e^{\sigma t} u_i(\mathbf{x}), \theta = e^{\sigma t}\theta(\mathbf{x}), q_i = e^{\sigma t} q_i(\mathbf{x}), \pi = e^{\sigma t}\pi(\mathbf{x}).$$

For two stress free surfaces a normal mode analysis may be performed as in Gentile and Straughan (2024) and the Rayleigh number expressions for stationary and oscillatory convection in Cattaneo–Morro theory are reduced to

$$R^2_{stat} = \frac{\Lambda^3}{a^2}\, \frac{2}{[2-(1+\gamma)SgPr\Lambda]}, \tag{3.22}$$

and

$$R^2_{osc} = \frac{\Lambda^2}{a^2}\, \frac{2Pr^2}{Sg[2Pr+1+\gamma]} + \frac{\Lambda}{a^2}\, \frac{2Pr(Pr+1)}{Sg^2[2Pr+1+\gamma]}, \tag{3.23}$$

where R^2_{stat} and R^2_{osc} are the Rayleigh numbers for stationary and oscillatory convection, respectively. The critical Rayleigh numbers are found by minimizing (3.22) and (3.23) in a^2.

Expressions (3.22) and (3.23) hold with $\gamma = 0$ for Cattaneo–Fox convection, with $\gamma = 1$ for CCR theory, and with $\gamma = -1$ for Cattaneo–Christov convection *or* for CCJ convection. It is noteworthy that the linearized theory for Cattaneo–Christov convection yields Rayleigh numbers which coincide with those for Cattaneo–Christov–Jordan convection.

Gentile and Straughan (2024) observe that equation (3.22) is consistent with the possibility of convective fluid motion if the layer is heated from above ($R^2_{stat} < 0$) when the denominator is negative. They argue that for physically correct behaviour one therefore needs $\gamma \le -1$. These writers also observe that for $\gamma \le -1$ from equation (3.23) one may have R^2_{osc} negative, and so oscillatory convection with heating from above. They then present an argument where one may neglect the Cattaneo–Fox and

CCR theories since $\gamma = 0$ and 1, respectively. The Cattaneo–Christov theory has $\gamma = -1$ and this in the linear instability situation agrees exactly with CCJ theory. Gentile and Straughan (2024) observe that in this case

$$R^2_{stat} = \frac{\Lambda^3}{a^2}, \qquad R^2_{osc} = \frac{Pr}{Sg}\frac{\Lambda^2}{a^2} + \frac{(Pr+1)}{Sg^2}\frac{\Lambda}{a^2},$$

cf. Straughan (2010b), Straughan (2011a, pages 233–237). Hence for two stress free surfaces $Ra_{stat} = 27\pi^4/4$, $a^2_{stat} = \pi^2/2$, and a^2_{osc}, Ra_{osc} may be found in equations (8.39), (8.40) of Straughan (2011a). Oscillatory convection may only occur for Sg large enough. A detailed analysis of thermal convection with CCJ theory, with a variety of boundary conditions, is given by Bissell (2015).

3.4 Nonlinear energy stability theory

One theory which has been very useful to obtain bounds ensuring nonlinear stability in thermal convection is to use an energy method, cf. Straughan (2004). This approach has been adopted by Gentile and Straughan (2024) and they show that the classical approach to nonlinear energy stability fails for the Cattaneo systems outlined in this chapter. We now give details of their work.

Let V be a periodic cell for the convective motion. Thus, V is typically a hexagonal shape in the (x, y) plane which forms a three dimensional cell by coupling with $z \in (0, 1)$, as shown in figure 3.1 where ∂V refers to the boundary of V.

We give details of the energy method for each of the five systems introduced above. To develop an energy stability analysis for the CCJ system we multiply $(3.16)_1$ by u_i and integrate over the cell V. Next, multiply $(3.16)_3$ by θ and likewise integrate over V. Then multiply $(3.16)_4$ by q_i and integrate over V. We integrate by parts and use the boundary conditions on u_i and θ to obtain the "energy" identities

$$\frac{d}{dt}\frac{1}{2}\|\mathbf{u}\|^2 = -\|\nabla\mathbf{u}\|^2 + R(\theta, w), \tag{3.24}$$

and

$$\frac{d}{dt}\frac{Pr}{2}\|\theta\|^2 = R(\theta, w) - (q_{i,i}, \theta), \tag{3.25}$$

and

$$\frac{d}{dt}\frac{Sg}{2}\|\mathbf{q}\|^2 = -\|\mathbf{q}\|^2 - (\theta_{,i}, q_i). \tag{3.26}$$

The idea behind the energy method is to add (3.24), (3.25) and (3.26) to form an energy equation. This has form

$$\frac{dE}{dt} = RI - D, \tag{3.27}$$

where the energy function E is defined by

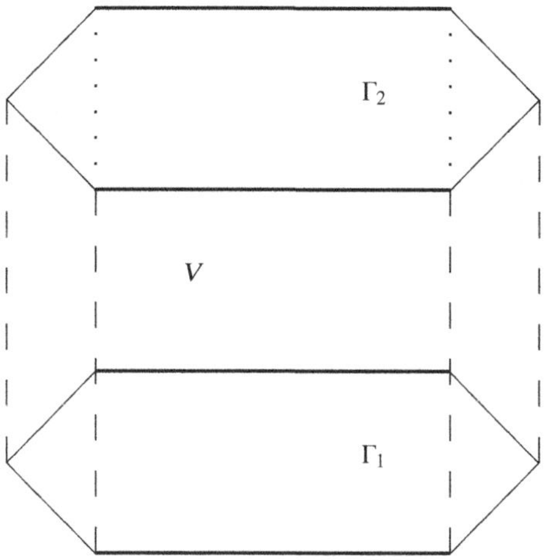

Fig. 3.1 A convection cell V. The surface Γ_1 is where ∂V lies in the plane $z = 0$ and Γ_2 is where ∂V lies in $z = 1$.

$$E(t) = \frac{1}{2}||\mathbf{u}||^2 + \frac{Pr}{2}||\theta||^2 + \frac{Sg}{2}||\mathbf{q}||^2, \tag{3.28}$$

the dissipation function D is defined by

$$D(t) = ||\nabla\mathbf{u}||^2 + ||\mathbf{q}||^2, \tag{3.29}$$

and the (indefinite) production term I is given by

$$I = 2(w, \theta). \tag{3.30}$$

In deriving the energy equation (3.27) we have used integration by parts and the boundary conditions to conclude that

$$-(\theta, q_{i,i}) - (\theta_{,i}, q_i) = 0.$$

Define now the quantity R_E by

$$\frac{1}{R_E} = \max_H \frac{I}{D}, \tag{3.31}$$

where H is the space of admissible solutions. From (3.27) one may deduce that

$$\frac{dE}{dt} = -D\Big(1 - \frac{R}{R_E}\Big), \tag{3.32}$$

and then if one had $R < R_E$ it is possible to use Poincaré's inequality and ensure $E(t)$ decays and nonlinear stability would follow. However, the key question is does the maximum in (3.31) exist?

Gentile and Straughan (2024) analyse the maximum problem for R_E, which is

$$\frac{1}{R_E} = \max_H \frac{2(\theta, w)}{\|\nabla \mathbf{u}\|^2 + \|\mathbf{q}\|^2}. \tag{3.33}$$

The space H is composed of the functions $\{\mathbf{u} \in (H^1(V))^3\}$ where $u_{i,i} = 0$ in V and u_i satisfies the boundary conditions, $u_i = 0$ on $z = 0, 1$, and $\{\theta \in H^1(V)\}$ with the boundary conditions $\theta = 0$ on $z = 0, 1$, and $\{\mathbf{q} \in L^2(V)\}$ together with u_i, θ, q_i satisfying periodicity in (x, y). Gentile and Straughan (2024) argue that since θ is effectively independent in the maximum of I/D then the maximum of this quantity cannot exist, since there is no term in θ in the denominator D. This is clearly a fundamental problem to the development of any stability threshold for the nonlinear theory from an energy theory point of view. One has to introduce a different strategy.

The problem just outlined for CCJ theory persists in all the other models we have considered. For Cattaneo–Fox theory one derives instead of (3.26)

$$\frac{d}{dt}\frac{Sg}{2}\|\mathbf{q}\|^2 = -(\theta_{,i}, q_i) - \|\mathbf{q}\|^2 + \frac{SgR}{2Pr}(u_{i,3}, q_i) - \frac{SgR}{2Pr}(w_{,i}, q_i). \tag{3.34}$$

Then, instead of (3.33) the quantity to be maximized is

$$\frac{I}{D} = \frac{2(\theta, w) + \dfrac{Sg}{2Pr}(u_{i,3}, q_i) - \dfrac{Sg}{2Pr}(w_{,i}, q_i)}{\|\nabla \mathbf{u}\|^2 + \|\mathbf{q}\|^2}. \tag{3.35}$$

The problem of achieving the maximum of this quotient persists in that there is still no θ present in the denominator, and one has to seek an alternative to standard energy theory.

When one employs Cattaneo–Christov theory, CCR theory, or the general Cattaneo–Morro theory with $\gamma \neq 0$, the nonexistence of the maximum of I/D

persists. However, in these cases a further complication arises due to the presence of a cubic nonlinear term. For example, for Cattaneo–Christov theory the identity which arises for q_i is

$$\frac{d}{dt}\frac{Sg}{2}\|\mathbf{q}\|^2 = -(\theta_{,i}, q_i) - \|\mathbf{q}\|^2 + \frac{SgR}{Pr}(u_{i,3}, q_i) + Sg\int_V d_{ij}q_iq_j\,dx. \tag{3.36}$$

where $d_{ij} = (u_{i,j} + u_{j,i})/2$. The last term is a cubic nonlinearity which would need special treatment in any case. However, the I/D term is here

$$\frac{I}{D} = \frac{2(\theta, w) + \dfrac{Sg}{Pr}(u_{i,3}, q_i)}{\|\nabla\mathbf{u}\|^2 + \|\mathbf{q}\|^2}. \tag{3.37}$$

Again, due to the fact that θ does not appear in the denominator one does not have existence of the maximum of I/D and a standard energy stability analysis does not appear possible.

For CCR theory the relevant energy identity for q_i is

$$\frac{d}{dt}\frac{Sg}{2}\|\mathbf{q}\|^2 = -(\theta_{,i}, q_i) - \|\mathbf{q}\|^2 - \frac{SgR}{Pr}(w_{,i}, q_i) - Sg\int_V d_{ij}q_iq_j\,dx. \tag{3.38}$$

The I/D term is

$$\frac{I}{D} = \frac{2(\theta, w) - \dfrac{Sg}{Pr}(w_{,i}, q_i)}{\|\nabla\mathbf{u}\|^2 + \|\mathbf{q}\|^2}. \tag{3.39}$$

Once again θ does not appear in the denominator D.

Finally, for Cattaneo–Morro theory the equation arising involving q_i is

$$\begin{aligned}\frac{d}{dt}\frac{Sg}{2}\|\mathbf{q}\|^2 =& \Big(\frac{1-\gamma}{2}\Big)\int_V u_{i,j}q_iq_j\,dx - \Big(\frac{1+\gamma}{2}\Big)\int_V u_{j,i}q_iq_j\,dx \\ &- \|\mathbf{q}\|^2 - (\theta_{,i}, q_i) \\ &+ \frac{SgR}{2Pr}(1-\gamma)(u_{i,3}, q_i) - \frac{SgR}{2Pr}(1+\gamma)(w_{,i}, q_i).\end{aligned} \tag{3.40}$$

The I/D term is

$$\frac{I}{D} = \frac{2(\theta, w) + \dfrac{Sg(1-\gamma)}{2Pr}(u_{i,3}, q_i) - \dfrac{Sg(1+\gamma)}{2Pr}(w_{,i}, q_i)}{\|\nabla\mathbf{u}\|^2 + \|\mathbf{q}\|^2}. \tag{3.41}$$

Thus, for all theories, the maximum of the production divided by the dissipation does not exist. Therefore a different approach is necessary. In the next chapter we consider such a new approach by modifying the equation of balance of energy and adopting a theory of Mariano (2017).

precisely. However, in [illegible] a further complication arises due to the presence [illegible]

[illegible]

Chapter 4
Cattaneo convection with Mariano theory

4.1 Balance of energy equation with a Mariano term

To remove the difficulty we saw in chapter 3 where energy stability theory failed due to the lack of dissipation in the temperature perturbation field, we turn to work of Mariano (2017) where the heat flux is split into two parts. Payne and Song (1997) added a Laplacian of the temperature field to the energy balance equation in Cattaneo theory and Straughan (2023c) followed their idea and explained it on the basis of work of Mariano (2017), by arguing that the extra flux term is connected to the microstructure of the material. This assertion about the microstructure has been justified rigorously by Mariano (2023). Morro (2024) is an interesting article which also considers the thermodynamic consistency of the Mariano theory and investigates compatibility with the second law of thermodynamics, within the theories of a mixture of viscous fluids or a mixture of a fluid and a solid, the latter having direct application to thermal convection in a fluid saturated porous medium.

We modify the energy balance equation which appears in the basic models (3.9) - (3.13) by employing an idea of Mariano (2017), where we split the heat flux into two parts. The energy balance equation in (3.9) - (3.13) is now replaced by

$$\frac{\partial T}{\partial t} + v_i \frac{\partial T}{\partial x_i} = -\frac{\partial Q_i}{\partial x_i} - \frac{\partial F_i}{\partial x_i}, \tag{4.1}$$

where the total heat flux is now $\mathbf{Q} + \mathbf{F}$. The F_i term is given by a Fourier law so

$$F_i = -\hat{\zeta} \frac{\partial T}{\partial x_i}, \tag{4.2}$$

where $\hat{\zeta} > 0$ is a constant. The flux Q_i is given by the appropriate equation in the models (3.9) - (3.13).

To interpret the significance of the two contributions to the heat flux Straughan (2023c) considers a rigid solid where $v_i \equiv 0$, and the system of equations for T becomes

B. Straughan, *Thermal Convection with a Cattaneo Flux Law*, Advances in Mechanics and Mathematics 53, https://doi.org/10.1007/978-3-031-80885-2_4

$$
\begin{aligned}
T_{,t} &= -Q_{i,i} - F_{i,i}, \\
F_i &= -\hat{\zeta} T_{,i}, \\
\tau Q_{i,t} + Q_i &= -\kappa T_{,i}.
\end{aligned} \tag{4.3}
$$

Then, equation $(4.3)_3$ may be integrated employing an integrating factor, and assuming fading memory of $T_{,i}$ one may show

$$
Q_i = -\frac{\kappa}{\tau} \int_{-\infty}^{t} \exp\left(-\frac{t-s}{\tau}\right) T_{,i}\, ds.
$$

Thus, the total heat flux is

$$
F_i + Q_i = -\hat{\zeta} T_{,i} - \frac{\kappa}{\tau} \int_{-\infty}^{t} \exp\left(-\frac{t-s}{\tau}\right) T_{,i}\, ds,
$$

which is a viscoelastic–like term for the temperature gradient plus a term emphasizing the instantaneous temperature gradient. As Straughan (2023c) points out, such a representation is common in viscoelasticity, cf. Boltzmann (1874), Miller (1978), or for a modern appreciation of the work of Boltzmann see Markowitz (1977).

With the addition of the Laplacian of the temperature field, the governing equations which replace (3.9) - (3.13) have form

$$
\begin{aligned}
&\frac{\partial v_i}{\partial t} + v_j \frac{\partial v_i}{\partial x_j} = -\frac{1}{\rho_0} \frac{\partial p}{\partial x_i} + \nu \Delta v_i + \alpha g k_i T, \\
&\frac{\partial v_i}{\partial x_i} = 0, \\
&\frac{\partial T}{\partial t} + v_i \frac{\partial T}{\partial x_i} = -\frac{\partial Q_i}{\partial x_i} + \hat{\zeta} \Delta T, \\
&\tau \frac{\mathcal{D} Q_i}{\mathcal{D} t} + Q_i = -\kappa \frac{\partial T}{\partial x_i}.
\end{aligned} \tag{4.4}
$$

In these equations the term $\mathcal{D}Q_i/\mathcal{D}t$ takes the form of a material derivative for CCJ theory, or it takes the form of an appropriate objective derivative, depending on which alternative theory is desired.

The basic state in the case of (4.4) remains the same as (3.14). We now proceed to analyse the stability and instability of this conduction solution.

Introducing perturbations as in section 3.2 the fully nonlinear, non-dimensional perturbation equations may be derived, cf. Gentile and Straughan (2024). If we let ζ be a non-dimensional form for $\hat{\zeta}$, then with CCJ theory the equations are

$$
\begin{aligned}
&u_{i,t} + u_j u_{i,j} = -\pi_{,i} + \Delta u_i + R k_i \theta, \\
&u_{i,i} = 0, \\
&Pr(\theta_{,t} + u_i \theta_{,i}) = Rw - q_{i,i} + \zeta \Delta \theta, \\
&Sg(q_{i,t} + u_j q_{i,j}) = -\theta_{,i} - q_i.
\end{aligned} \tag{4.5}
$$

For Cattaneo–Morro theory the relevant equations are

$$
\begin{aligned}
& u_{i,t} + u_j u_{i,j} = -\pi_{,i} + \Delta u_i + Rk_i\theta, \\
& u_{i,i} = 0, \\
& Pr(\theta_{,t} + u_i\theta_{,i}) = Rw - q_{i,i} + \zeta\Delta\theta, \\
& Sg(q_{i,t} + u_j q_{i,j} - \omega_{ij}q_j + \gamma d_{ij}q_j) = -\theta_{,i} - q_i + \frac{SgR}{Pr}\,\omega_{i3} - \gamma\frac{Sg}{R}\,d_{i3},
\end{aligned}
\tag{4.6}
$$

where

$$
d_{ij} = \frac{1}{2}(u_{i,j} + u_{j,i}), \qquad \omega_{ij} = \frac{1}{2}(u_{i,j} - u_{j,i}).
$$

Of course, with $\gamma = 0$ the equations of Cattaneo–Fox theory are obtained in the case of a Mariano extra flux, whereas $\gamma = -1$ yields Cattaneo–Christov theory, $\gamma = 1$ gives CCR theory, and an arbitrary constant γ leads to Cattaneo–Morro theory, although we stress these are when the Mariano extra flux theory is employed.

4.2 Energy stability for CCJ–Mariano theory

Gentile and Straughan (2024) develop a linear instability analysis for (4.5) by employing a normal mode technique, cf. Barletta (2019, 2021). They show that the Rayleigh number for stationary convection is given by

$$
R^2_{stat} = (\zeta + 1)\frac{\Lambda^3}{a^2}, \tag{4.7}
$$

whereas the oscillatory convection Rayleigh number is derived from

$$
R^2_{osc} = \zeta\frac{\Lambda^3}{a^2} + \left[\frac{\zeta + Pr^2 + \zeta^2 + 2\zeta Pr}{Sg(\zeta + Pr)}\right]\frac{\Lambda^2}{a^2} + \left[\frac{Pr^2 + (1+\zeta)Pr}{Sg^2(\zeta + Pr)}\right]\frac{\Lambda}{a^2}. \tag{4.8}
$$

The critical Rayleigh and wave numbers are found by minimizing R^2_{stat} and R^2_{osc} in a^2. In this way they show that for stationary convection with two stress free surfaces

$$
R^2_{stat} = \frac{27\pi^4}{4}(\zeta + 1), \qquad a^2_{stat} = \frac{\pi^2}{2}. \tag{4.9}
$$

Gentile and Straughan (2024) find the minimum of the oscillatory convection boundary minimizing (4.8) in a^2 by a numerical technique, and numerical results are given in that work.

Gentile and Straughan (2024) develop a nonlinear energy stability analysis for equations (4.5). They arrive at an energy equation

$$
\frac{dE}{dt} = RI - D
$$

where

$$
E = \frac{1}{2}\|\mathbf{u}\|^2 + \frac{Pr}{2}\|\theta\|^2 + \frac{Sg}{2}\|\mathbf{q}\|^2,
$$

the production term I is

$$I = 2(w, \theta)$$

and the dissipation D is

$$D = \|\nabla \mathbf{u}\|^2 + \zeta \|\nabla \theta\|^2 + \|\mathbf{q}\|^2.$$

Define R_E by

$$R_E = \max_H \frac{I}{D} \tag{4.10}$$

where H is the space of admissible solutions. Then

$$\frac{dE}{dt} \le -D\Big(1 - \frac{R}{R_E}\Big)$$

and so if $R < R_E$ we find with $b = 1 - R/R_E > 0$, and use of Poincaré's inequality

$$\frac{dE}{dt} = -bk_1 E\,,$$

for a constant $k_1 > 0$. Hence, E decays rapidly and one obtains global stability.

Gentile and Straughan (2024) show that the Euler–Lagrange equations for (4.10) are

$$\begin{aligned} &\Delta u_i + R_E \theta k_i = \varphi_{,i}\,, \\ &u_{i,i} = 0, \\ &\zeta \Delta \theta + R_E w = 0, \end{aligned} \tag{4.11}$$

where φ is a Lagrange multiplier. Gentile and Straughan (2024) solve these equations for two free surfaces and find

$$R_E^2 = \zeta \frac{\Lambda^3}{a^2}\,, \tag{4.12}$$

whence the critical nonlinear energy Rayleigh number is

$$R_E^2 = \frac{27\pi^4 \zeta}{4}\,. \tag{4.13}$$

Gentile and Straughan (2024) observe that

$$R_{stat}^2 \ge R_E^2 \qquad \text{and} \qquad R_{osc}^2 \ge R_E^2\,,$$

inequalities which ought to hold. They note that the boundaries yielded by R_{stat}^2 and R_{osc}^2 represent linear instability thresholds, whereas those of R_E^2 represent a threshold for global nonlinear stability.

4.3 Energy stability for Cattaneo–Fox–Mariano theory

Gentile and Straughan (2024) do not consider an energy analysis explicitly for the Cattaneo–Fox theory when the Mariano heat flux term is present, due to the fact that the linear theory predicts instability while heating from above. It is important to stress that the Cattaneo–Fox–Mariano theory yields a perfectly good global nonlinear energy stability analysis which we include here. For other objective derivatives rather than the co-rotational one the situation is not so easy, as we see in section 4.4. It is of interest to note that Morro (2024) employs the co-rotational derivative in his development of general rate type (Cattaneo like) models for a mixture of fluids, or for a mixture of a fluid and a solid, the latter having much application to fluid flow in porous media.

In the interests of clarity we recall the perturbation equations for Cattaneo–Fox–Mariano theory here from (4.6),

$$\begin{aligned} &u_{i,t}+u_ju_{i,j}=-\pi_{,i}+\Delta u_i+Rk_i\theta,\\ &u_{i,i}=0,\\ &Pr(\theta_{,t}+u_i\theta_{,i})=Rw-q_{i,i}+\zeta\Delta\theta,\\ &Sg(q_{i,t}+u_jq_{i,j}-\omega_{ij}q_j)=-\theta_{,i}-q_i+\frac{SgR}{Pr}\,\omega_{i3}\,. \end{aligned} \tag{4.14}$$

The boundary conditions are

$$u_i=0,\qquad \theta=0,\qquad \text{on } z=0,1, \tag{4.15}$$

with u_i, θ, q_i, π satisfying a plane tiling periodicity in the x, y plane.

To develop an energy stability analysis multiply $(4.14)_1$ by u_i, $(4.14)_3$ by θ, $(4.14)_4$ by q_i, and integrate each over a period cell V. After integration by parts and use of the boundary conditions, one obtains

$$\frac{d}{dt}\frac{1}{2}\|\mathbf{u}\|^2=-\|\nabla\mathbf{u}\|^2+R(\theta,w), \tag{4.16}$$

and

$$\frac{d}{dt}\frac{Pr}{2}\|\theta\|^2=-\zeta\|\nabla\theta\|^2+R(\theta,w)-(q_{i,i},\theta), \tag{4.17}$$

together with

$$\frac{d}{dt}\frac{Sg}{2}\|\mathbf{q}\|^2=-\zeta\|\mathbf{q}\|^2+R(\theta,w)-(q_i,\theta_{,i})+\frac{SgR}{Pr}(\omega_{i3},q_i). \tag{4.18}$$

Introduce a constant coupling parameter $\lambda>0$ and then by addition one finds

$$\frac{dE}{dt}=R(1+\lambda)(\theta,w)-\lambda\|\nabla\mathbf{u}\|^2-\zeta\|\nabla\theta\|^2-\|\mathbf{q}\|^2+\frac{SgR}{Pr}\,(\omega_{i3},q_i)\,, \tag{4.19}$$

where

$$E = \frac{\lambda}{2}||\mathbf{u}||^2 + \frac{Pr}{2}||\theta||^2 + \frac{Sg}{2}||\mathbf{q}||^2. \tag{4.20}$$

From the arithmetic - geometric mean inequality

$$\begin{aligned}(\omega_{i3}, q_i) &\leq \frac{1}{2\epsilon}||\mathbf{q}||^2 + \frac{\epsilon}{2}\int_V \omega_{i3}\omega_{i3}dx, \\ &\leq \frac{1}{2\epsilon}||\mathbf{q}||^2 + \frac{\epsilon}{2}\int_V \omega_{ij}\omega_{ij}dx, \\ &= \frac{1}{2\epsilon}||\mathbf{q}||^2 + \frac{\epsilon}{4}||\nabla\mathbf{u}||^2, \end{aligned} \tag{4.21}$$

where $\epsilon > 0$ is a constant at our disposal. Let δ be a small positive number, $0 < \delta << 1$. Then, for fixed R put

$$k = \frac{Sg^2R^2}{8Pr^2(1-\delta)} > 0.$$

We suppose Sg is small and we now choose $\lambda = 1 + 2k$. Next, employ (4.21) in (4.19) to derive with $\epsilon = SgR/2Pr(1-\delta)$,

$$\frac{dE}{dt} \leq RI - D - \delta||\mathbf{q}||^2,$$

where

$$I = 2(1+k)(\theta, w), \qquad D = \zeta||\nabla\theta||^2 + (1+k)||\nabla\mathbf{u}||^2.$$

Define

$$\frac{1}{R_E} = \max_H \frac{I}{D}, \tag{4.22}$$

where H is the space consisting of those functions which satisfy $u_i \in H^1(V)$, $\theta \in H^1(V)$, $u_{i,i} = 0$, together with the boundary conditions (4.15). Then it follows that

$$\frac{dE}{dt} \leq -D\Big(1 - \frac{R}{R_E}\Big) - \delta||\mathbf{q}||^2. \tag{4.23}$$

Suppose $R < R_E$, then put $b = 1 - R/R_E > 0$. Then (4.23) is

$$\frac{dE}{dt} \leq -bD - \delta||\mathbf{q}||^2. \tag{4.24}$$

Upon using Poincaré's inequality in (4.24) one shows there is a positive constant c such that

$$\frac{dE}{dt} \leq -cE. \tag{4.25}$$

Thus, if $R < R_E$, E decreases rapidly in time and global nonlinear stability follows.

The Euler–Lagrange equations from (4.22) are

$$
\begin{aligned}
&\Delta u_i + R_E \theta k_i = \omega_{,i} , \\
&u_{i,i} = 0, \\
&\zeta \Delta \theta + (1 + k) R_E w = 0,
\end{aligned}
\tag{4.26}
$$

where ω is a Lagrange multiplier. From these equations one deduces for two stress free surfaces

$$
R_E^2 = \frac{\Lambda^3}{a^2} \frac{\zeta}{(1 + k)} . \tag{4.27}
$$

Thus, the energy stability limit is

$$
R_E^2 = \frac{27\pi^4 \zeta}{4(1 + k)} .
$$

4.4 Energy stability for Cattaneo–Morro–Mariano theory

In the context of a general objective derivative, Gentile and Straughan (2024) analysed nonlinear energy stability only for the Cattaneo–Christov case including a Mariano heat flux. They concentrated on this case due to the fact that the theories involving the other objective derivatives all suffer from the drawback that linear instability theory allows for the unphysical behaviour of convective instability when the fluid layer is heated from above. The work of Gentile and Straughan (2024) shows that for the Cattaneo–Christov case one must add a Laplacian of the heat flux into the Cattaneo equation. Such an addition is possible, and may be desirable, see Ván and Fülöp (2012, equation (5)), Mariano (2017, equation (1.71)). Gentile and Straughan (2024) showed that an energy stability analysis may be developed when the period cell V is two-dimensional. For the three-dimensional situation they were not able to proceed and they had to include a Kelvin–Voigt term into the momentum equation and deal with a Navier–Stokes–Voigt theory. For both the two and three dimensional problems Gentile and Straughan (2024) obtained a conditional nonlinear energy stability analysis, i.e. one where the initial data must be sufficiently small in a precise sense.

The method employed by Gentile and Straughan (2024) will work for the general Cattaneo–Morro theory. We outline the approach here for Cattaneo–Christov theory and for CCR theory. The Cattaneo–Morro system with a Mariano flux, (4.6), is modified to include a heat flux diffusion and the equations then become

$$
\begin{aligned}
&u_{i,t} + u_j u_{i,j} = -\pi_{,i} + \Delta u_i + R k_i \theta, \\
&u_{i,i} = 0, \\
&Pr(\theta_{,t} + u_i \theta_{,i}) = R w - q_{i,i} + \zeta \Delta \theta, \\
&Sg(q_{i,t} + u_j q_{i,j} - \omega_{ij} q_j + \gamma d_{ij} q_j) = -\theta_{,i} - q_i \\
&\qquad\qquad + \frac{SgR}{Pr} \omega_{i3} - \gamma \frac{Sg}{R} d_{i3} + \epsilon \Delta q_i ,
\end{aligned}
\tag{4.28}
$$

where $\epsilon > 0$ is a constant. The term involving ϵ represents the diffusion of heat flux, cf. Ván and Fülöp (2012, equation 5), Mariano (2017, equation 1.71).

In the interests of clarity we display the analogous systems of equations for Cattaneo–Christov and CCR theory which arise from (4.28). For Cattaneo–Christov theory they are

$$\begin{aligned} &u_{i,t} + u_j u_{i,j} = -\pi_{,i} + \Delta u_i + Rk_i\theta, \\ &u_{i,i} = 0, \\ &Pr(\theta_{,t} + u_i\theta_{,i}) = Rw - q_{i,i} + \zeta\Delta\theta, \\ &Sg(q_{i,t} + u_j q_{i,j} - u_{i,j}q_j) = -\theta_{,i} - q_i + \frac{SgR}{Pr}\, u_{i,3} + \epsilon\Delta q_i\,, \end{aligned} \tag{4.29}$$

whereas for CCR theory they become

$$\begin{aligned} &u_{i,t} + u_j u_{i,j} = -\pi_{,i} + \Delta u_i + Rk_i\theta, \\ &u_{i,i} = 0, \\ &Pr(\theta_{,t} + u_i\theta_{,i}) = Rw - q_{i,i} + \zeta\Delta\theta, \\ &Sg(q_{i,t} + u_j q_{i,j} + u_{j,i}q_j) = -\theta_{,i} - q_i - \frac{SgR}{Pr}\, w_{,i} + \epsilon\Delta q_i\,. \end{aligned} \tag{4.30}$$

In all cases, i.e. for all $\gamma \neq 0$, it would appear necessary to divide the energy stability analysis into a two-dimensional one and a three-dimensional one.

4.4.1 Energy stability in two space dimensions

At this point we let the period cell V be either a two or three dimensional one. To develop an energy stability analysis one commences by multiplying $(4.29)_1$ (or equivalently $(4.30)_1$) by u_i and one then integrates the result over the domain V. Next multiply $(4.29)_3$, or $(4.30)_1$ by θ, and integrate over V. After integration by parts and use of the boundary conditions one obtains the identities

$$\frac{d}{dt}\frac{1}{2}\|\mathbf{u}\|^2 = -\|\nabla\mathbf{u}\|^2 + R(\theta, w), \tag{4.31}$$

and

$$\frac{d}{dt}\frac{Pr}{2}\|\theta\|^2 = -\zeta\|\nabla\theta\|^2 + R(\theta, w) - (q_{i,i}, \theta). \tag{4.32}$$

At this point the energy identity one obtains from the heat flux equation depends on which objective derivative is used. The idea is to multiply $(4.29)_3$ by q_i when dealing with Cattaneo–Christov theory, whereas one multiplies $(4.30)_3$ by q_i when dealing with CCR theory. One integrates the result in either case over V and after some integration by parts and use of the boundary conditions one may arrive at the following results. For Cattaneo–Christov theory one finds

$$\begin{aligned}\frac{d}{dt}\frac{Sg}{2}\|\mathbf{q}\|^2 =&Sg\int_V d_{ij}q_iq_j\,dx + \frac{RSg}{Pr}(u_{i,3},q_i)\\ &-\epsilon\|\nabla\mathbf{q}\|^2-(q_i,\theta_{,i})-\|\mathbf{q}\|^2.\end{aligned}\tag{4.33}$$

For CCR theory one instead derives

$$\begin{aligned}\frac{d}{dt}\frac{Sg}{2}\|\mathbf{q}\|^2 =&-Sg\int_V d_{ij}q_iq_j\,dx - \frac{RSg}{Pr}(w_{,i},q_i)\\ &-\epsilon\|\nabla\mathbf{q}\|^2-(q_i,\theta_{,i})-\|\mathbf{q}\|^2.\end{aligned}\tag{4.34}$$

The next step is to add (4.31), (4.32) and (4.33) for Cattaneo–Christov theory or to add (4.31), (4.32) and (4.34) for CCR theory. In either case the sum $-(q_{i,i},\theta)-(q_i,\theta_{,i})$ is zero as can be seen using integration by parts and the boundary conditions.

The resulting energy equations which arise by forming the sum as indicated are, for Cattaneo–Christov theory,

$$\begin{aligned}\frac{dE}{dt} =&RI+Sg\int_V d_{ij}q_iq_j\,dx + \frac{RSg}{Pr}(u_{i,3},q_i)\\ &-\|\nabla\mathbf{u}\|^2-\epsilon\|\nabla\mathbf{q}\|^2-\|\mathbf{q}\|^2-\zeta\|\nabla\theta\|^2,\end{aligned}\tag{4.35}$$

while that for CCR theory is

$$\begin{aligned}\frac{dE}{dt} =&RI-Sg\int_V d_{ij}q_iq_j\,dx - \frac{RSg}{Pr}(w_{,i},q_i)\\ &-\|\nabla\mathbf{u}\|^2-\epsilon\|\nabla\mathbf{q}\|^2-\|\mathbf{q}\|^2-\zeta\|\nabla\theta\|^2.\end{aligned}\tag{4.36}$$

In these equations the energy functional E, and the production term I, have form

$$E=\frac{1}{2}\|\mathbf{u}\|^2+\frac{Pr}{2}\|\theta\|^2+\frac{Sg}{2}\|\mathbf{q}\|^2,\tag{4.37}$$

and

$$I=2(\theta,w).\tag{4.38}$$

Now observe that from the arithmetic - geometric mean inequality

$$(u_{i,3},q_i)\le\frac{1}{2\delta}\|\nabla\mathbf{u}\|^2+\frac{\delta}{2}\|\mathbf{q}\|^2,\tag{4.39}$$

and

$$(w_{,i},q_i)\le\frac{1}{2\delta}\|\nabla\mathbf{u}\|^2+\frac{\delta}{2}\|\mathbf{q}\|^2,\tag{4.40}$$

for a positive constant δ at our disposal. Thus,

$$\frac{RSg}{Pr}(u_{i,3},q_i)\le\frac{1}{2}\|\nabla\mathbf{u}\|^2+\frac{R^2Sg^2}{2Pr^2}\|\mathbf{q}\|^2,\tag{4.41}$$

and

$$\frac{RSg}{Pr}(w_{,i},q_i) \le \frac{1}{2}\,\|\nabla\mathbf{u}\|^2 + \frac{R^2Sg^2}{2Pr^2}\,\|\mathbf{q}\|^2. \tag{4.42}$$

Now suppose Sg is small enough that

$$Sg^2 \le \frac{2Pr^2}{R^2}. \tag{4.43}$$

Then from either (4.35) or (4.36) we may derive the energy inequality

$$\frac{dE}{dt} \le RI - D_1 + Sg\int_V |q_iq_jd_{i,j}|\,dx - \epsilon\|\nabla\mathbf{q}\|^2, \tag{4.44}$$

where

$$D_1 = \frac{1}{2}\|\nabla\mathbf{u}\|^2 + \zeta\|\nabla\theta\|^2. \tag{4.45}$$

The procedure is now as in Gentile and Straughan (2024). Define

$$\frac{1}{R_E} = \max_H \frac{I}{D_1}, \tag{4.46}$$

where H is the space of admissible solutions for u_i and θ, i.e. $\mathbf{u} \in (H^1(V))^3$, $\theta \in H^1(V)$, $u_{i,i} = 0$, and u_i and θ satisfy the appropriate boundary conditions. Define D by

$$D = D_1 + \epsilon\|\nabla\mathbf{q}\|^2,$$

and then from (4.44) one obtains

$$\frac{dE}{dt} \le -D\Big(1 - \frac{R}{R_E}\Big) + Sg\int_V |q_iq_jd_{i,j}|\,dx\,. \tag{4.47}$$

The maximum problem (4.46) is solved in Gentile and Straughan (2024).

It is at this point where we restrict V to be a two-dimensional domain. In two space dimensions one employs the Sobolev inequality (1.23). Suppose $R < R_E$ and let $b = 1 - R/R_E > 0$. From (4.44) one sees that

$$\begin{aligned}\frac{dE}{dt} &\le -\,bD + \frac{Sg}{\sqrt{2}}\,\|\mathbf{q}\|_4^2\,\|\nabla\mathbf{u}\|\,,\\ &\le -\,bD + \frac{Sg\,c_1}{\sqrt{2}}\,\|\mathbf{q}\|\,\|\nabla\mathbf{q}\|\,\|\nabla\mathbf{u}\|\,,\\ &\le -\,(b - kE^{1/2})D,\end{aligned} \tag{4.48}$$

where $k = c_1\sqrt{2Sg/\epsilon}$. When $E^{1/2}(0) < b/k$, Gentile and Straughan (2024) demonstrate that $E \to 0$ at least exponentially, and so conditional nonlinear stability is obtained with the nonlinear energy stability threshold R_E.

4.4.2 Energy stability in three space dimensions

In three dimensions the analysis leading to (4.48) does not work because the Sobolev inequality (1.23) must be replaced by (1.24). With E and D as in section 4.4.1, the Sobolev inequality in three space dimensions does not lead to inequality (4.48). Gentile and Straughan (2024) suggested one way to remedy this by adding a Kelvin–Voigt regularizing term to equations (4.29). This is equivalent to working with a Navier–Stokes–Voigt fluid rather than a Navier–Stokes one.

The Navier–Stokes–Voigt perturbation equations coupled with a Mariano extra flux are for Cattaneo–Christov theory,

$$\begin{aligned}
&u_{i,t} + u_j u_{i,j} - \lambda\Delta u_{i,t} = -\pi_{,i} + \Delta u_i + R\theta k_i\,,\\
&u_{i,i} = 0,\\
&Pr(\theta_{,t} + u_i\theta_{,i}) = Rw - q_{i,i} + \zeta\Delta\theta,\\
&Sg(q_{i,t} + u_j q_{i,j} - u_{i,j}q_j) = \frac{RSg}{Pr}\,u_{i,3} - q_i - \theta_{,i} + \epsilon\Delta q_i,
\end{aligned} \tag{4.49}$$

where $\lambda > 0$ is here a constant. For CCR theory the analogous system of equations is

$$\begin{aligned}
&u_{i,t} + u_j u_{i,j} - \lambda\Delta u_{i,t} = -\pi_{,i} + \Delta u_i + R\theta k_i\,,\\
&u_{i,i} = 0,\\
&Pr(\theta_{,t} + u_i\theta_{,i}) = Rw - q_{i,i} + \zeta\Delta\theta,\\
&Sg(q_{i,t} + u_j q_{i,j} + u_{j,i}q_j) = -\frac{RSg}{Pr}\,w_{,i} - q_i - \theta_{,i} + \epsilon\Delta q_i\,.
\end{aligned} \tag{4.50}$$

For either system of equations (4.49) or (4.50) the analysis of section 4.4.1 is repeated and one may arrive at the inequality (4.47) where D is as in section 4.4.1 but where the energy function now has the form

$$E(t) = \frac{1}{2}\|\mathbf{u}\|^2 + \frac{\lambda}{2}\|\nabla\mathbf{u}\|^2 + \frac{Pr}{2}\|\theta\|^2 + \frac{Sg}{2}\|\mathbf{q}\|^2. \tag{4.51}$$

The key difference with section 4.4.1 is the presence of the term $\lambda\|\nabla\mathbf{u}\|^2/2$ in $E(t)$. The first inequality in (4.48) is still valid, namely

$$\frac{dE}{dt} \le -bD + \frac{Sg}{\sqrt{2}}\,\|\mathbf{q}\|_4^2\,\|\nabla\mathbf{u}\|\,. \tag{4.52}$$

One now employs the Sobolev inequality (1.24) in inequality (4.52) to see that

$$\begin{aligned}
\frac{dE}{dt} &\le -\,bD + \frac{Sg\,c_2}{\sqrt{2}}\,\|\mathbf{q}\|^{1/2}\,\|\nabla\mathbf{q}\|^{3/2}\,\|\nabla\mathbf{u}\|\,,\\
&\le -\,(b - kE^{1/2})D,
\end{aligned} \tag{4.53}$$

where $k = 2^{3/4}/(Sg\lambda\epsilon^3)^{1/4}$. From (4.53) one requires $R < R_E$ and $E^{1/2}(0) < b/k$ and then one may employ a continuity argument, cf. Straughan (2004, pages 14–16), to deduce the energy $E(t)$ tends to zero exponentially as $t \rightarrow \infty$.

Hence, conditional nonlinear stability for the Navier–Stokes–Voigt system of equations with a Mariano flux is established for a solution to the Cattaneo–Christov or CCR models.

Chapter 5
Guyer–Krumhansl effects

The present chapter is effectively derived from Straughan (2023c) who developed a novel model for thermal convection using Brinkman–Darcy–Kelvin–Voigt theory but he allowed the heat flux to be of Maxwell–Cattaneo type. The novelties of that work involve employing Guyer–Krumhansl theory in conjunction with Maxwell–Cattaneo theory, cf. Jou et al. (2010), Ván et al. (2017); splitting the heat flux into an instantaneous Fourier part and a memory part depending on the temperature gradient, using an extra flux idea of Mariano (2017); and allowing the heat flux equation to be objective, which necessarily introduces other terms, cf. Morro (2018, 2022). Straughan (2023c) argues that incorporation of Guyer–Krumhansl theory is pertinent since Ván et al. (2017), Fülöp et al. (2018), suggest that this theory may be more relevant than Fourier theory, even at room temperatures. A complete stability analysis for the Bénard problem for the Brinkman–Darcy–Kelvin–Voigt equations is presented by Straughan (2023c). He derived a detailed linear instability analysis for both stationary and oscillatory convection and he included a fully nonlinear energy stability analysis. He observed that such energy stability analyses for systems incorporating Maxwell–Cattaneo effects have previously proved difficult, cf. Straughan (2004, page 192).

We explain that even though the Guyer–Krumhansl terms appear as diffusion terms in the differential equation for the heat flux these terms lead to destabilization of the fluid layer when stationary convection occurs. This effect is somewhat surprising since diffusion usually stabilizes the onset of convective fluid motion, and the phenomenon whereby the Guyer–Krumhansl terms are more unstable was found by Franchi and Straughan (1994).

5.1 Instability in Cattaneo–Fox theory with Guyer–Krumhansl terms

We now review work of Franchi and Straughan (1994) who worked with the Cattaneo–Fox model without a Mariano extra flux, and so we begin with equa-

B. Straughan, *Thermal Convection with a Cattaneo Flux Law*, Advances in Mechanics and Mathematics 53, https://doi.org/10.1007/978-3-031-80885-2_5

tions (3.10) but we include two terms in the equation for the heat flux, these being the Guyer–Krumhansl contributions, cf. Ván et al. (2017). We change the notation of Franchi and Straughan (1994) to be consistent with that of chapter 3 and the rest of the book.

The basic equations for thermal convection in a Cattaneo–Fox model incorporating Guyer–Krumhansl terms are

$$
\begin{aligned}
&\frac{\partial v_i}{\partial t} + v_j \frac{\partial v_i}{\partial x_j} = -\frac{1}{\rho_0}\frac{\partial p}{\partial x_i} + \nu \Delta v_i + \alpha g k_i T,\\
&\frac{\partial v_i}{\partial x_i} = 0,\\
&\frac{\partial T}{\partial t} + v_i \frac{\partial T}{\partial x_i} = -\frac{\partial Q_i}{\partial x_i},\\
&\tau\left(\frac{\partial Q_i}{\partial t} + v_j \frac{\partial Q_i}{\partial x_j} + \frac{Q_j}{2}\left[\frac{\partial v_j}{\partial x_i} - \frac{\partial v_i}{\partial x_j}\right]\right) + Q_i = -k\frac{\partial T}{\partial x_i} + \hat{\xi}_1 \Delta Q_i + \hat{\xi}_2 Q_{k,ik}\,,
\end{aligned}
\tag{5.1}
$$

where $\hat{\xi}_1$ and $\hat{\xi}_2$ denote the terms associated with the Guyer–Krumhansl theory.

The basic solution to (5.1) in whose stability we are interested is given by the stationary solution (3.14). Perturbation variables are introduced as in chapter 3 and these are non-dimensionalized with the scalings of that chapter together with non-dimensional versions of the Guyer–Krumhansl coefficients ξ_1 and ξ_2. Then, the non-dimensional perturbation equations are

$$
\begin{aligned}
&u_{i,t} + u_j u_{i,j} = -\pi_{,i} + \Delta u_i + R k_i \theta,\\
&u_{i,i} = 0,\\
&Pr(\theta_{,t} + u_i \theta_{,i}) = Rw - q_{i,i}\,,\\
&Sg\Big(q_{i,t} + u_j q_{i,j} - \frac{1}{2}u_{i,j}q_j + \frac{1}{2}u_{j,i}q_j\Big) =\\
&\qquad -\theta_{,i} - q_i + \frac{SgR}{2Pr}u_{i,3} - \frac{SgR}{2Pr}w_{,i} + \xi_1 \Delta q_i + \xi_2 q_{k,ik}\,.
\end{aligned}
\tag{5.2}
$$

These equations hold on the layer $\mathbb{R}^2 \times (0,1)$ for $t > 0$. The boundary conditions we consider here are that the solution is periodic in x, y and

$$
u_i = 0, \quad \theta = 0, \quad \mathbf{q} \times \mathbf{n} = 0, \tag{5.3}
$$

on the planes $z = 0$ and 1. It follows from the condition $\mathbf{q} \times \mathbf{n} = 0$ on the planes $z = 0, 1$, that $q_1 = 0$ and $q_2 = 0$ on $z = 0, 1$. Further applying $(5.2)_3$ on the boundaries $z = 0, 1$, it then follows that $q_{3,3} = 0$ on $z = 0, 1$.

To describe the results of Franchi and Straughan (1994) we linearize (5.2) and introduce the time dependence like $e^{\sigma t}$. After removing the pressure π, and taking the divergence of $(5.2)_4$, one may then show that w, θ and $\Gamma = q_{i,i}$ satisfy the system of equations

$$\begin{aligned} -\sigma\Delta w &= -R\Delta^*\theta - \Delta^2 w, \\ \sigma Pr\theta &= Rw - \Gamma, \\ \sigma Sg\Gamma &= -\Gamma - \Delta\theta - \frac{SgR}{2Pr}\Delta w + \epsilon\Delta\Gamma, \end{aligned} \tag{5.4}$$

where $\epsilon = \xi_1 + \xi_2$. System (5.4) together with the boundary conditions $w = \theta = \Gamma = 0$ on $z = 0, 1$ and the periodicity of the solution is an eigenvalue problem for the growth rate σ.

For two stress free surfaces one employs a normal mode analysis to derive from (5.4) the equation

$$\begin{aligned} \sigma^3 PrSg\Lambda + \sigma^2\big[PrSg\Lambda^2 + Pr\Lambda(1+\epsilon\Lambda)\big] + \sigma\big[Pr\Lambda^2(1+\epsilon\Lambda) + \Lambda^2\big] \\ + \Lambda^3 = R^2a^2\Big(\sigma Sg + 1 + \epsilon\Lambda - \frac{Sg\Lambda}{2Pr}\Big). \end{aligned} \tag{5.5}$$

The stationary convection threshold is found immediately from (5.5) as

$$R^2_{stat} = \frac{\Lambda^3}{a^2}\,\frac{1}{(1+\epsilon_1\Lambda)}, \tag{5.6}$$

where $\epsilon_1 = \epsilon - Sg/2Pr$.

To derive the oscillatory convection threshold one puts $\sigma = i\omega$ and takes the real and imaginary parts of (5.5) to obtain the two equations

$$-\omega^2\big[PrSg\Lambda^2 + Pr\Lambda(1+\epsilon\Lambda)\big] + \Lambda^3 = R^2a^2\Big(1 + \epsilon\Lambda - \frac{Sg\Lambda}{2Pr}\Big), \tag{5.7}$$

and

$$-\omega^2 PrSg\Lambda + Pr\Lambda^2(1+\epsilon\Lambda) + \Lambda^2 = R^2a^2Sg\,. \tag{5.8}$$

The variable ω is eliminated between (5.7) and (5.8) and in this way the oscillatory convection instability threshold is found to be given by

$$R^2_{osc} = \Big(\frac{2Pr}{1+2Pr}\Big)\Big[\frac{Pr}{Sg}\frac{\Lambda^2}{a^2}(1+\epsilon\Lambda) + \frac{Pr}{Sg^2}\frac{\Lambda}{a^2}(1+\epsilon\Lambda)^2 + \frac{\Lambda}{a^2}\Big(\frac{1+\epsilon\Lambda}{Sg^2}\Big)\Big]. \tag{5.9}$$

The critical values of Ra and a^2 are found by minimizing (5.6) and (5.9) in a^2.

The behaviour of the critical Rayleigh number is like that shown in figure 8.1 of Straughan (2011a). For Sg small enough stationary convection occurs and then at a critical value of Sg oscillatory convection becomes the dominant mode of instability. Clearly, from (5.9) oscillatory convection may only commence for Sg large enough.

When $\epsilon > Sg/2Pr$, in the region where stationary convection occurs Franchi and Straughan (1994, page 372) note that, ... "*the addition of the Guyer–Krumhansl theory yields lower critical Rayleigh numbers than the classical theory and so convection should commence sooner, i.e. the layer is less stable.*" Table 1 of Franchi and Straughan (1994, page 373) yields the critical values of Ra for $\epsilon = 10^{-4}, 10^{-3}, 0.01, 0.1, 1, 10$ and100 and it is clearly seen that as ϵ_1 increases the critical Rayleigh number decreases rapidly (monotonically) from the value $Ra = 656.539$ when $\epsilon = 10^{-4}$ to $Ra = 0.395$ when $\epsilon_1 = 100$. This demonstrates that the Guyer–Krumhansl effect coupled with the Cattaneo theory for the heat flux behaves in a counter intuitive manner with the Guyer–Krumhansl diffusion–like terms leading to destabilization of the stationary solution as opposed to stabilization which diffusion is usually responsible for.

We have highlighted the above effect of the Guyer–Krumhansl terms in the context of the original work of Franchi and Straughan (1994) who employed Cattaneo–Fox theory. However, it is also true in other Cattaneo theories such as Cattaneo–Christov or CCJ, as we now show.

5.2 Instability in Cattaneo–Christov theory with Guyer–Krumhansl terms

We now analyse the analogous problem to that of section 5.1 but we employ a Cattaneo–Christov model without a Mariano extra flux, and so we begin with equations (3.11) but we include two terms in the equation for the heat flux, these being the Guyer–Krumhansl contributions, cf. Ván et al. (2017).

The basic equations for thermal convection in a Cattaneo–Christov model incorporating Guyer–Krumhansl terms are

$$\begin{aligned}
&\frac{\partial v_i}{\partial t} + v_j\frac{\partial v_i}{\partial x_j} = -\frac{1}{\rho_0}\frac{\partial p}{\partial x_i} + \nu\Delta v_i + \alpha g k_i T,\\
&\frac{\partial v_i}{\partial x_i} = 0,\\
&\frac{\partial T}{\partial t} + v_i\frac{\partial T}{\partial x_i} = -\frac{\partial Q_i}{\partial x_i},\\
&\tau\Big(\frac{\partial Q_i}{\partial t} + v_j\frac{\partial Q_i}{\partial x_j} - Q_j\frac{\partial v_i}{\partial x_j}\Big) + Q_i = -k\frac{\partial T}{\partial x_i} + \hat{\xi}_1\Delta Q_i + \hat{\xi}_2 Q_{k,ik}\,.
\end{aligned} \tag{5.10}$$

The basic solution to (5.10) in whose stability we are interested is given by the stationary solution (3.14). Perturbation variables are introduced as in chapter 3 and these are non-dimensionalized with the scalings of that chapter, cf. section 5.1. The non-dimensional perturbation equations in this case are

$$\begin{aligned}
&u_{i,t} + u_j u_{i,j} = -\pi_{,i} + \Delta u_i + Rk_i\theta,\\
&u_{i,i} = 0,\\
&Pr(\theta_{,t} + u_i\theta_{,i}) = Rw - q_{i,i}\,,\\
&Sg(q_{i,t} + u_j q_{i,j} - u_{i,j}q_j) = -\theta_{,i} - q_i + \frac{SgR}{Pr}u_{i,3} + \xi_1\Delta q_i + \xi_2 q_{k,ik}\,.
\end{aligned} \tag{5.11}$$

These equations hold on the layer $\mathbb{R}^2 \times (0,1)$ for $t > 0$. The boundary conditions we consider here are that the solution is periodic in x, y and

$$u_i = 0, \quad \theta = 0, \quad \mathbf{q} \times \mathbf{n} = 0, \tag{5.12}$$

on the planes $z = 0$ and 1. Again, we note that $\mathbf{q} \times \mathbf{n} = 0$ on $z = 0, 1$, leads to the boundary conditions

$$q_1 = q_2 = 0, \quad q_{3,3} = 0, \qquad \text{on } z = 0, 1.$$

Next linearize (5.11) and introduce the time dependence like $e^{\sigma t}$. After removing the pressure π, and taking the divergence of $(5.11)_4$, one may then show that w, θ and $\Gamma = q_{i,i}$ satisfy the system of equations

$$\begin{aligned} -\sigma \Delta w &= -R\Delta^* \theta - \Delta^2 w, \\ \sigma Pr \theta &= Rw - \Gamma, \\ \sigma Sg \Gamma &= -\Gamma - \Delta\theta + \epsilon \Delta \Gamma, \end{aligned} \tag{5.13}$$

where $\epsilon = \xi_1 + \xi_2$. System (5.13) together with the boundary conditions $w = \theta = \Gamma = 0$ on $z = 0, 1$ and the periodicity of the solution is an eigenvalue problem for the growth rate σ.

For two stress free surfaces one employs a normal mode analysis to derive from (5.13) the equation

$$\begin{aligned} \sigma^3 PrSg\Lambda + \sigma^2\big[PrSg\Lambda^2 + Pr\Lambda(1+\epsilon\Lambda)\big] + \sigma\big[Pr\Lambda^2(1+\epsilon\Lambda) + \Lambda^2\big] \\ + \Lambda^3 = R^2a^2(\sigma Sg + 1 + \epsilon\Lambda). \end{aligned} \tag{5.14}$$

The stationary convection threshold is found from (5.14) as

$$R^2_{stat} = \frac{\Lambda^3}{a^2} \frac{1}{(1+\epsilon\Lambda)}\,. \tag{5.15}$$

To derive the oscillatory convection threshold one puts $\sigma = i\omega$ and takes the real and imaginary parts of (5.14) to obtain the two equations

$$-\omega^2\big[PrSg\Lambda^2 + Pr\Lambda(1+\epsilon\Lambda)\big] + \Lambda^3 = R^2a^2(1+\epsilon\Lambda), \tag{5.16}$$

and

$$-\omega^2 PrSg\Lambda + Pr\Lambda^2(1+\epsilon\Lambda) + \Lambda^2 = R^2a^2Sg\,. \tag{5.17}$$

The variable ω is eliminated between (5.16) and (5.17) to find the oscillatory convection instability threshold as

$$R^2_{osc} = \frac{Pr}{Sg}\frac{\Lambda^2}{a^2}(1+\epsilon\Lambda) + \frac{Pr}{Sg^2}\frac{\Lambda}{a^2}(1+\epsilon\Lambda)^2 + \frac{\Lambda}{a^2}\frac{1+\epsilon\Lambda}{Sg^2}\,. \tag{5.18}$$

The critical values of Ra and a^2 are found by minimizing (5.15) and (5.18) in a^2.

The behaviour of the critical Rayleigh number is like that shown in figure 8.2 of Straughan (2011a, page 237). For Sg small enough stationary convection occurs and then at a critical value of Sg oscillatory convection becomes the dominant mode of instability. It is evident from (5.18) that oscillatory convection may only commence for Sg large enough.

We do not include numerical details of the minimization of (5.15) or (5.18), but we believe that from the structure of (5.15), the Guyer–Krumhansl terms will again be destabilizing the fluid layer. One may formally take the derivative of R^2_{stat} with respect to ϵ to see that $\partial R^2_{stat}/\partial\epsilon < 0$. Thus, it is likely that the minimum values of R^2_{stat} with respect to a^2 will show the instability behaviour found in section 5.1. Thus, we expect that the Guyer–Krumhansl effect coupled with the Cattaneo–Christov theory (or for the CCJ model) for the heat flux will behave in a counter intuitive manner with the Guyer–Krumhansl diffusion–like terms leading to destabilization of the stationary solution as opposed to stabilization which diffusion is usually responsible for.

5.3 Stability in Cattaneo–Christov–Mariano theory with Guyer–Krumhansl terms

In this section we essentially add the Guyer–Krumhansl effect to the work previously considered in chapter 4, although we concentrate on a Cattaneo–Christov model and incorporate a Mariano split heat flux. To present the model we let $v_i(\mathbf{x},t), p(\mathbf{x},t), T(\mathbf{x},t)$ and $Q_i(\mathbf{x},t)$ denote the velocity, pressure, temperature, and heat flux at position $\mathbf{x}$ and time t. The model of thermal convection considered here consists of a balance of momentum equation,

$$v_{i,t} + v_j v_{i,j} = -\frac{1}{\rho} p_{,i} + \nu\Delta v_i + \alpha g k_i T\,, \tag{5.19}$$

a balance of mass equation

$$v_{i,i} = 0, \tag{5.20}$$

a balance of energy equation

$$T_{,t} + v_i T_{,i} = -Q_{i,i} + \zeta\Delta T, \tag{5.21}$$

and a Cattaneo - Christov like equation for the heat flux of form

$$\tau(Q_{i,t} + v_j Q_{i,j} - v_{i,j}Q_j) = -Q_i - \kappa T_{,i} + \hat{\xi}_1\Delta Q_i + \hat{\xi}_2 Q_{k,ik}\,. \tag{5.22}$$

In these equations ρ is the constant density of the fluid, ν is the kinematic viscosity, α is the thermal expansion coefficient, g is gravity, $\mathbf{k} = (0,0,1)$, ζ is a positive constant, τ is the relaxation time, κ is the thermal diffusivity, and $\hat{\xi}_1, \hat{\xi}_2$ are the Guyer–Krumhansl coefficients.

Equations (5.19) - (5.22) are defined on the horizontal layer $\{(x,y)\in\mathbb{R}^2\}\times\{z\in(0,d)\}$ for $t>0$. The boundary conditions are that

$$v_i = 0,\ z=0,d; \quad T=T_L\,,\ z=0; \quad T=T_U\,,\ z=d; \quad Q_i=0,\ z=0,d; \tag{5.23}$$

$0<T_U<T_L$, with $T_L\,,T_U$ constants.

We choose a Guyer–Krumhansl theory in (5.22), cf. Ván et al. (2017), Fülöp et al. (2018), Capriz et al. (2021), and we observe in particular that Ván et al. (2017) is an interesting article containing experimental and theoretical work which suggests Guyer–Krumhansl theory may be more accurate than Fourier or traditional Maxwell–Cattaneo theory, even for temperatures typical of laboratory experiments.

The steady conduction solution to equations (5.19), (5.20), (5.21), (5.22) and (5.23) in whose stability we are interested has form

$$\bar{v}_i \equiv 0, \qquad \bar{T} = T_L - \beta z, \qquad \bar{\mathbf{Q}} = (0,0,\kappa\beta), \tag{5.24}$$

where β is the temperature gradient,

$$\beta = \frac{T_L - T_U}{d} > 0.$$

The steady pressure is a quadratic function of z determined from (5.19).

To analyse the instability/stability of the steady solution (5.24) we define perturbation variables u_i, θ, q_i, π by

$$v_i = \bar{v}_i + u_i, \qquad T = \bar{T} + \theta, \qquad Q_i = \bar{Q}_i + q_i, \qquad p = \bar{p} + \pi.$$

These expressions are substituted into equations (5.19), (5.20), (5.21) and (5.22) and are non-dimensionalized with the scalings

$$\begin{aligned}
&x_i = x_i^* d, \qquad t = \mathcal{T} t^*, \qquad \mathcal{T} = \frac{d^2}{\nu},\\
&u_i = u_i^* U, \qquad \theta = \theta^* T^\sharp, \qquad q_i = q_i^* Q^\sharp, \qquad \pi = \pi^* P\\
&U = \frac{\nu}{d}, \qquad Pr = \frac{\nu}{\kappa}, \qquad T^\sharp = U\sqrt{\frac{\beta\nu}{\alpha g\kappa}} \qquad Q^\sharp = \frac{T^\sharp\kappa}{d}\\
&\xi_1 = \frac{\hat{\xi}_1 Q^\sharp}{d\kappa T^\sharp}, \qquad \xi_2 = \frac{\hat{\xi}_2 Q^\sharp}{d\kappa T^\sharp},\\
&P = \frac{\rho\nu U}{d}, \qquad Sg = \frac{\tau\nu}{d^2}.
\end{aligned}$$

The Rayleigh number Ra is defined as

$$Ra \equiv R^2 = \frac{\beta g\alpha d^4}{\nu\kappa}.$$

The non-dimensional perturbation equations, where the *s are dropped, may be shown to be

$$
\begin{aligned}
&u_{i,t} + u_j u_{i,j} = -\pi_{,i} + \Delta u_i + R\theta k_i\,,\\
&u_{i,i} = 0,\\
&Pr(\theta_{,t} + u_i\theta_{,i}) = Rw - q_{i,i} + \zeta\Delta\theta\,,\\
&Sg(q_{i,t} + u_j q_{i,j} - u_{i,j} q_j) = RPrSg\, u_{i,3} - q_i - \theta_{,i} + \xi_1\Delta q_i + \xi_2 q_{k,ki}\,,
\end{aligned}
\tag{5.25}
$$

where (5.25) hold on $\mathbb{R}^2 \times (0,1) \times \{t > 0\}$, with $w \equiv u_3$.

As Straughan (2023c) observes, care must be taken with the boundary conditions when Guyer–Krumhansl terms are present. The non-dimensional perturbation boundary conditions normally used are

$$
u_i = 0, \quad z = 0,1; \qquad \theta = 0, \quad z = 0,1; \qquad q_i = 0, \quad z = 0,1; \tag{5.26}
$$

or

$$
\begin{aligned}
&u_i = 0, \quad z = 0,1; \qquad \theta = 0, \quad z = 0,1;\\
&q_1 = q_2 = 0, \quad z = 0,1; \qquad q_{3,3} = 0, \quad z = 0,1;
\end{aligned}
\tag{5.27}
$$

together with horizontal periodicity of the solution. The period cell of the solution is denoted by V. When one employs (5.27) one additionally requires $\int_V q_3 dx = 0$, to exclude constant vertical heat flux perturbations.

In this section we concentrate on stress free boundary conditions with (5.27). To develop a linear instability analysis for (5.25) and (5.27) we drop the quadratic terms and introduce a time dependence like

$$
u_i = e^{\sigma t}u_i(\mathbf{x}), \quad \theta = e^{\sigma t}\theta(\mathbf{x}), \quad q_i = e^{\sigma t}q_i(\mathbf{x}), \quad \pi = e^{\sigma t}\pi(\mathbf{x}).
$$

As Straughan (2023c) shows, the condition $n_j\sigma_{ij} = 0$ on $z = 0,1$, reduces to

$$
u_{,z} = v_{,z} = 0, \qquad \text{on} \quad z = 0,1,
$$

and then with the continuity equation one sees that

$$
w_{,zz} = 0, \qquad \text{on} \quad z = 0,1.
$$

We also require

$$
w = 0,\ \theta = 0,\ q_1 = 0,\ q_2 = 0,\ q_{3,3} = 0, \qquad \text{on } z = 0,1.
$$

For linear instability one then reduces equations (5.25) to

$$
\begin{aligned}
&-\sigma\Delta w + \Delta^2 w + R\Delta^*\theta = 0,\\
&Pr\sigma\theta - \zeta\Delta\theta - Rw + \Gamma = 0,\\
&(Sg\sigma + 1)\Gamma + \Delta\theta - \epsilon\Delta\Gamma = 0,
\end{aligned}
\tag{5.28}
$$

where $\Delta^* = \partial^2/\partial x^2 + \partial^2/\partial y^2$, the variable Γ is introduced by $\Gamma = q_{i,i}$, and $\epsilon = \xi_1 + \xi_2$. For stress free boundary conditions one may show from (5.28) that w, θ and Γ may be represented by sin series of form $w = \sum_{n=1}^{\infty} w_n \sin n\pi z$, with a similar form for θ, Γ.

The linear instability boundary one finds from (5.28) is given by solving the determinant equation

$$\begin{vmatrix} \sigma_n\Lambda_n + \Lambda_n^2 & -Ra^2 & 0 \\ -R & \sigma_n Pr + \zeta\Lambda_n & 1 \\ 0 & -\Lambda_n & \sigma_n Sg + 1 + \epsilon\Lambda_n \end{vmatrix} = 0,$$

where $\Lambda_n = n^2\pi^2 + a^2$, a being the wavenumber, and $n = 1, 2, \ldots$ This determinant equation leads to the equation

$$\begin{aligned} &\sigma_n^3 Sg Pr \Lambda_n + \sigma_n^2\big[Sg\Lambda_n^2(Pr + \zeta) + Pr\Lambda_n(1 + \epsilon\Lambda_n)\big] \\ &\quad + \sigma_n\big[\Lambda_n^2(Pr + \zeta)(1 + \epsilon\Lambda_n) + Sg\zeta\Lambda_n^3 + \Lambda_n^2\big] \\ &\quad + \zeta\Lambda_n^3(1 + \epsilon\Lambda_n) + \Lambda_n^3 = R^2a^2\sigma_n Sg + R^2a^2(1 + \epsilon\Lambda_n). \end{aligned} \tag{5.29}$$

The stationary convection threshold ($\sigma_n = 0$) is then found as

$$R_{stat}^2 = \zeta\frac{\Lambda_n^3}{a^2} + \frac{\Lambda_n^3}{a^2(1 + \epsilon\Lambda_n)}\,. \tag{5.30}$$

By taking the derivative in (5.30) we see that $\partial R_{stat}^2/\partial n^2 \geq 0$ and so we may limit attention to $n = 1$, whence

$$R_{stat}^2 = \zeta\frac{\Lambda^3}{a^2} + \frac{\Lambda^3}{a^2(1 + \epsilon\Lambda)}\,, \tag{5.31}$$

where $\Lambda = \pi^2 + a^2$. The stationary convection critical values of the Rayleigh number $Ra = R^2$ are found by minimizing R_{stat}^2 in a^2.

To find the oscillatory convection instability threshold one puts $\sigma = i\omega$ in (5.29). Actually one puts $\sigma_n = i\omega_n$ but it is found afterward that $n = 1$ yields the critical value. Then we take the real and imaginary parts of the equation which follows from (5.29) and this yields the two equations

$$-\omega^2\big[Sg\Lambda^2(Pr + \zeta) + Pr\Lambda(1 + \epsilon\Lambda)\big] + \Lambda^3(1 + \zeta + \zeta\epsilon\Lambda) = R^2a^2(1 + \epsilon\Lambda), \tag{5.32}$$

and

$$-\omega^2 Sg Pr\Lambda + \Lambda^2(Pr + \zeta)(1 + \epsilon\Lambda) + Sg\zeta\Lambda^3 + \Lambda^2 = R^2a^2 Sg. \tag{5.33}$$

The variable ω is eliminated from (5.32) and (5.33) to find

$$
\begin{aligned}
R^2_{osc} =&\zeta\frac{\Lambda^3}{a^2}+\frac{\Lambda^2}{a^2}\frac{(Pr+\zeta)(1+\epsilon\Lambda)}{Sg}+\frac{\Lambda}{a^2}\frac{Pr(1+\epsilon\Lambda)^2}{Sg^2}\\
&+\frac{\Lambda^2}{a^2}\frac{\zeta}{(Pr+\zeta)Sg}+\frac{\Lambda(1+\epsilon\Lambda)Pr}{a^2Sg^2(Pr+\zeta)}\,.
\end{aligned} \tag{5.34}
$$

The critical oscillatory convection values are determined by minimizing R^2_{osc} in a^2.

5.3.1 Nonlinear energy stability

One way to obtain an unconditional nonlinear stability result for a solution to (5.25) is to take the divergence of $(5.25)_4$ and work with the system

$$
\begin{aligned}
&u_{i,t}+u_ju_{i,j}=-\pi_{,i}+\Delta u_i+R\theta k_i\,,\\
&u_{i,i}=0,\\
&Pr(\theta_{,t}+u_i\theta_{,i})=Rw-\Gamma+\zeta\Delta\theta\,,\\
&Sg(\Gamma_{,t}+u_i\Gamma_{,i})=-\Gamma-\Delta\theta+\epsilon\Delta\Gamma\,.
\end{aligned} \tag{5.35}
$$

We then multiply $(5.35)_1$ by u_i, $(5.35)_3$ by θ, and $(5.35)_4$ by Γ, and integrate each over a period cell V. After some integration by parts and use of the boundary conditions $(5.26)_2$ one may arrive at the identities

$$
\begin{aligned}
&\frac{d}{dt}\frac{1}{2}\|\mathbf{u}\|^2=-\|\nabla\mathbf{u}\|^2+R(\theta,w),\\
&\frac{d}{dt}\frac{Pr}{2}\|\theta\|^2=R(\theta,w)-(\Gamma,\theta)-\zeta\|\nabla\theta\|^2,\\
&\frac{d}{dt}\frac{Sg}{2}\|\Gamma\|^2=-\|\Gamma\|^2+(\nabla\theta,\nabla\Gamma)-\epsilon\|\nabla\Gamma\|^2.
\end{aligned} \tag{5.36}
$$

Now form the combination $\lambda_2\times(5.36)_1+\lambda_1\times(5.36)_3+(5.36)_4$, for positive coupling parameters λ_1,λ_2 at our disposal. This results in the following equation

$$
\begin{aligned}
&\frac{d}{dt}\Big(\frac{\lambda_2}{2}\|\mathbf{u}\|^2+\frac{\lambda_1 Pr}{2}\|\theta\|^2+\frac{Sg}{2}\|\Gamma\|^2\Big)\\
&=-\lambda_2\|\nabla\mathbf{u}\|^2+R(\lambda_1+\lambda_2)(\theta,w)-\lambda_1(\Gamma,\theta)\\
&\quad-\lambda_1\zeta\|\nabla\theta\|^2-\|\Gamma\|^2+(\nabla\theta,\nabla\Gamma)-\epsilon\|\nabla\Gamma\|^2.
\end{aligned} \tag{5.37}
$$

There are various approaches possible at this point. We outline one such approach. Let D_1 be defined by

$$
\begin{aligned}
D_1 =&\epsilon\|\nabla\Gamma\|^2+\lambda_1\zeta\|\nabla\theta\|^2-(\nabla\theta,\nabla\Gamma)\\
\ge&\Big(\epsilon-\frac{\alpha}{2}\Big)\|\nabla\Gamma\|^2+\Big(\lambda_1\zeta-\frac{1}{2\alpha}\Big)\|\nabla\theta\|^2,
\end{aligned}
$$

where in the last line we have used the arithmetic - geometric mean inequality for $\alpha > 0$. We now require that the constants $\epsilon, \lambda_1, \zeta$ are such that

$$\epsilon > \frac{\alpha}{2} \qquad \text{and} \qquad \lambda_1 \zeta > \frac{1}{2\alpha} .$$

For example, we might choose $\alpha = \epsilon$ and then this requires

$$\lambda_1 \geq \frac{1}{2\zeta}\Big(\epsilon + \frac{1}{\epsilon}\Big) . \tag{5.38}$$

Suppose this is now so and then from (5.37) we may derive the following energy inequality

$$\frac{dE}{dt} \leq -D + I, \tag{5.39}$$

where

$$\begin{aligned} E(t) &= \frac{\lambda_2}{2}\|\mathbf{u}\|^2 + \frac{\lambda_1 Pr}{2}\|\theta\|^2 + \frac{Sg}{2}\|\Gamma\|^2 , \\ I(t) &= (\lambda_1 + \lambda_2)R(w,\theta) - \lambda_1(\Gamma,\theta) , \\ D(t) &= \|\Gamma\|^2 + \frac{\epsilon}{2}\|\nabla\Gamma\|^2 + \frac{\epsilon}{2}\|\nabla\theta\|^2 + \lambda_2\|\nabla\mathbf{u}\|^2 . \end{aligned}$$

From (5.39) we see that

$$\frac{dE}{dt} \leq -D\Big(1 - \frac{I}{D}\Big), \tag{5.40}$$

and then if

$$\max_H \frac{I}{D} < 1, \tag{5.41}$$

where H is the space of admissible solutions, we may employ Poincaré's inequality in (5.40) to find a positive constant k such that

$$\frac{dE}{dt} \leq -kE .$$

Thus E decays rapidly and nonlinear stability follows. The nonlinear stability restriction is (5.41) and the Euler–Lagrange equations arising from this are

$$\begin{aligned} &(\lambda_1 + \lambda_2)R\theta k_i + 2\lambda_2 \Delta u_i = \omega_{,i} , \\ &u_{i,i} = 0, \\ &(\lambda_1 + \lambda_2)Rw - \lambda_1 \Gamma + \epsilon\Delta\theta = 0, \\ &-\lambda_1\theta - 2\Gamma + \epsilon\Delta\Gamma = 0, \end{aligned}$$

where ω is a Lagrange multiplier.

The Euler–Lagrange system of equations yields the nonlinear stability threshold equation as

$$R^2\frac{(\lambda_1+\lambda_2)^2}{2\lambda_2}\,a^2(\epsilon\Lambda+2)=\epsilon\Lambda^3(\epsilon\Lambda+2)-\lambda_1\Lambda^2.$$

Then, with λ_1 given by (5.38) we maximize the expression for R^2 in terms of λ_2 to find $\lambda_2=\lambda_1$. This yields a nonlinear energy stability boundary as

$$R_{en}^2=\frac{1}{2\lambda_2(2+\epsilon\Lambda)}\Big(\epsilon^2\frac{\Lambda^4}{a^2}+2\epsilon\frac{\Lambda^3}{a^2}-\lambda_1^2\frac{\Lambda^2}{a^2}\Big)\,. \tag{5.42}$$

The critical energy stability values are found from (5.42) by minimizing R_{en}^2 in a^2, accounting for the constraint (5.38).

Remark. If we had chosen to employ a CCJ theory rather than a Cattaneo–Christov one in section 5.3.1 then an unconditional nonlinear stability analysis does not follow in the same way. Indeed with an analogous CCJ theory equation $(5.35)_4$ should be replaced by

$$Sg(\Gamma_{,t}+u_i\Gamma_{,i}+u_{j,i}q_{i,j})=-\Gamma-\Delta\theta+\epsilon\Delta\Gamma.$$

The third term on the left of this equation leads to problems if one derives an equation analogous to $(5.36)_3$, since a cubic term of form $Sg\int_V u_{j,i}q_{i,j}\Gamma\,dx$ needs to be handled.

5.3.2 Numerical results

We now report numerical results for the linear instability critical Rayleigh number and wavenumber based on (5.31) and (5.34). These results are contained in figures 5.1 - 5.4 and tables 5.1 - 5.4.

Tables 5.1 and 5.2 report critical threshold instability values where $Pr=0.1,\zeta=0.1$ and ϵ varies from 0.1 to 1. Tables 5.3 and 5.4 report critical threshold instability values where $Pr=0.1,\epsilon=0.1$ and ζ varies from 0.1 to 1. The typical behaviour of the Rayleigh number, Ra, as Sg increases is shown in figure 5.1 for various values of ϵ and in figure 5.2 for several values of ζ. The behaviour in all cases is qualitatively similar. When Sg is small stationary convection dominates and at a certain value of Sg oscillatory convection becomes the dominant mechanism for Sg larger, cf. Straughan (2011a, figure 8.2). The transition values from stationary convection to oscillatory convection are given in detail in tables 5.1 - 5.4. We have computed the critical Rayleigh number for Sg becoming very large and it appears that Ra approaches the asymptotic value expected from (5.34). That is $Ra\to\zeta\Lambda^3/a^2$, so $a_{crit}^2\to\pi^2/2$ and $Ra\to 27\zeta\pi^4/4$.

The variation of the critical Rayleigh number shown in figures 5.1 and 5.2 is very different depending on whether ϵ is varied or ζ is changed. With the variation of the Mariano coefficient ζ, Ra increases as ζ increases for both stationary and oscillatory convection. This is to be expected since ζ is a diffusion coefficient for the temperature field. However, the Guyer–Krumhansl parameter ϵ may also be thought of as a diffusion term for the heat flux. In this case the behaviour shown in figure

5.1 and in figures 5.3 and 5.4 is interesting. Figure 5.4 shows that as ϵ is increased the oscillatory convection threshold increases meaning the Guyer–Krumhansl term is acting as a typical diffusion and is stabilizing the layer. Figure 5.3, on the other hand, shows that the Guyer–Krumhansl effect is acting counter intuitively and as ϵ increases the layer becomes less stable with stationary convection. The second order nature in time of the temperature part of the problem would appear to be influencing how diffusion acts. This is a case which is somewhat reminiscent of the effect of diffusion in Turing instabilities.

We do not include explicit figures like 5.3 and 5.4 for the variation in ζ since they would both increase as ζ increases. The novelty appears to be connected with the Guyer–Krumhansl effect in the evolution equation for the heat flux.

We do not discuss in detail wavenumber behaviour, but the critical values for both stationary and oscillatory convection are given in tables 5.1 - 5.4. We also do not include details of the global energy stability results from (5.42) since for the range of parameters we have investigated they are much lower than the linear instability ones.

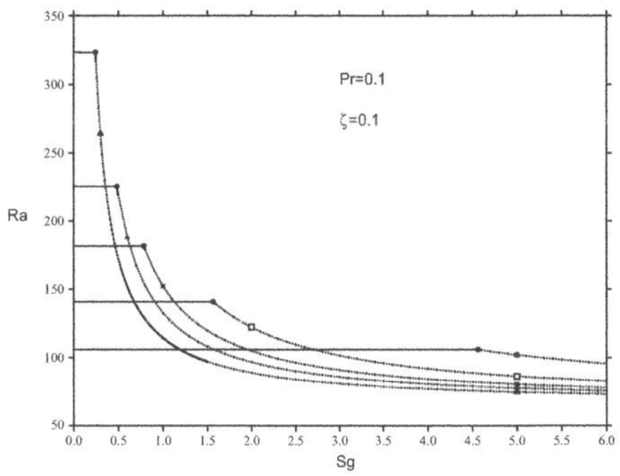

Fig. 5.1 Ra vs. Sg with $Pr = 0, 1, \zeta = 0.1$. The horizontal straight line on each curve is the stationary convection threshold whereas the dashed curves show the oscillatory convection branches. The curves shown are for $\epsilon = 0.1$, △; $\epsilon = 0.2$, ▲; $\epsilon = 0.3$, ×; $\epsilon = 0.5$, □; $\epsilon = 1$, •.

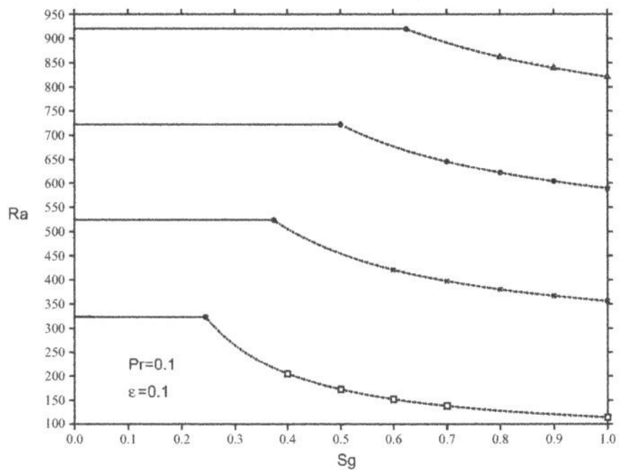

Fig. 5.2 Ra vs. Sg with $Pr = 0, 1, \epsilon = 0.1$. The horizontal straight line on each curve is the stationary convection threshold whereas the dashed curves show the oscillatory convection branches. The curves shown are for $\zeta = 0.1$, □; $\zeta = 0.4$, ×; $\zeta = 0.7$, •; $\epsilon = 1$, △.

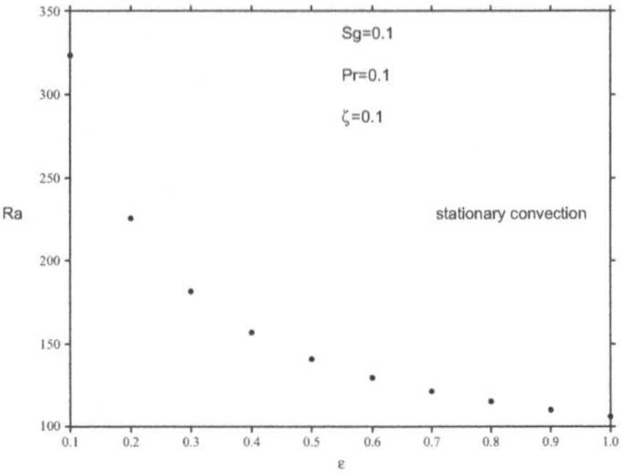

Fig. 5.3 Ra vs. ϵ with $Pr = 0, 1$, $\zeta = 0.1$. Stationary convection values given when $Sg = 0.1$.

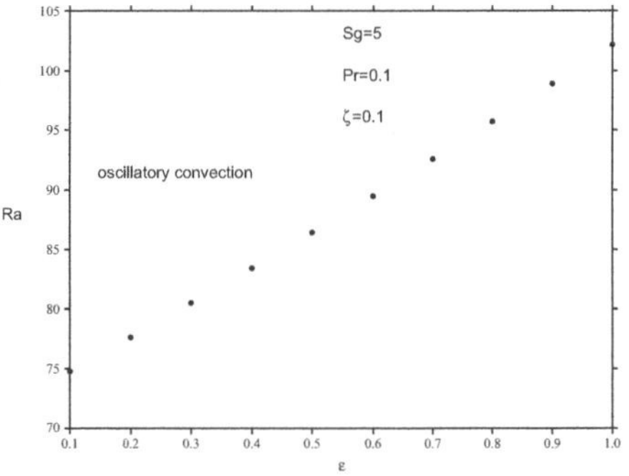

Fig. 5.4 Ra vs. ϵ with $Pr = 0, 1$, $\zeta = 0.1$. Oscillatory convection values given when $Sg = 5$.

5.4 Mariano and Guyer–Krumhansl effects in Darcy porous convection

5.4.1 Zero fluid inertia

We commence by considering a saturated porous medium of Darcy type where we neglect fluid inertia. The porous medium is saturated with an incompressible fluid and the basic equations follow from Straughan (2011a, pages 238-241). In fact, Straughan (2011a, pages 238-241) analyses instability in a saturated porous medium of Darcy type, but Guyer–Krumhansl and Mariano terms are not considered there. If we include both Mariano and Guyer–Krumhansl terms then for a Cattaneo–Christov theory the governing equations are

$$
\begin{aligned}
&0 = -\frac{1}{\rho}\frac{\partial p}{\partial x_i} + \alpha g k_i T - \frac{\mu}{\rho K}\, v_i\,,\\
&v_{i,i} = 0,\\
&\frac{1}{M}\frac{\partial T}{\partial t} + v_i\frac{\partial T}{\partial x_i} = -\frac{\partial Q_i}{\partial x_i} + \hat{\zeta}\Delta T,\\
&\tau\frac{\partial Q_i}{\partial t} + \tau_f\Big(v_j\frac{\partial Q_i}{\partial x_j} - Q_j\frac{\partial v_i}{\partial x_j}\Big) = -Q_i - \kappa\frac{\partial T}{\partial x_i} + \hat{\xi}_1\Delta Q_i + \hat{\xi}_2 Q_{j,ij}\,.
\end{aligned}
\tag{5.43}
$$

Table 5.1 Critical Rayleigh and wave numbers; $Pr = 0.1$, $\zeta = 0.1$.

Ra_{stat}	Ra_{osc}	a^2_{stat}	a^2_{osc}	ϵ	Sg
323.396	977.154	6.54	8.47	0.1	0.1
323.396	323.472	6.543	7.298	0.1	0.245
323.396	322.106	6.543	7.293	0.1	0.246
323.396	74.774	6.543	5.16	0.1	5
323.396	70.219	6.543	5.05	0.1	10
225.290	1612.55	6.74	7.56	0.2	0.1
225.290	225.421	6.745	6.306	0.2	0.484
225.290	225.007	6.745	6.303	0.2	0.485
225.290	77.608	6.745	5.15	0.2	5
225.290	71.585	6.745	5.05	0.2	10
181.614	2380.23	6.67	7.04	0.3	0.1
181.614	199.133	6.675	5.860	0.3	0.784
181.614	181.708	6.675	5.859	0.3	0.785
181.614	181.527	6.675	5.15	0.3	5
181.614	71.585	6.675	5.05	0.3	10
156.889	3280.08	6.55	6.70	0.4	0.1
156.889	156.975	6.553	5.610	0.4	1.144
156.889	156.880	6.553	5.610	0.4	1.145
156.889	83.432	6.553	5.14	0.4	5
156.889	74.356	6.553	5.04	0.4	10
140.953	4311.94	6.43	6.46	0.5	0.1
140.953	140.961	6.430	5.453	0.5	1.565
140.953	140.904	6.430	5.453	0.5	1.566
140.953	86.423	6.430	5.14	0.5	5
140.953	75.761	6.430	5.04	0.5	10

Here ρ, p, T, v_i and Q_i are the constant fluid density, the pressure, the temperature, the pore-averaged velocity, and the heat flux. The coefficients $\alpha, g, \mu, K, \kappa, \tau$ and τ_f are the fluid thermal expansion coefficient, gravity, the dynamic viscosity of the fluid, the permeability of the porous medium, the thermal conductivity of the porous medium, the relaxation coefficient for the porous medium, and the relaxation coefficient for the fluid. We suppose that $\tau = \phi\tau_f + (1-\phi)\tau_s$, where τ_f and τ_s are the relaxation times for the fluid and solid skeleton, and ϕ is the porosity. The coefficient M is

$$M = \frac{(\rho c_p)_f}{(\rho c)_m}, \qquad \text{where} \qquad (\rho c)_m = \phi(\rho c_p)_f + (1-\phi)(\rho c)_s,$$

where f, s, m denote fluid, solid, and porous medium, and c is the specific heat, c_p being the specific heat of the fluid at constant pressure. The parameters $\hat{\xi}_1$ $\hat{\xi}_2$ are the Guyer–Krumhansl coefficients.

Equations (5.43) hold on the layer $\mathbb{R}^2 \times \{z \in (0,d)\}$ for $t > 0$. The solution to (5.43) is subject to the boundary conditions

Table 5.2 Critical Rayleigh and wave numbers; $Pr = 0.1$, $\zeta = 0.1$.

Ra_{stat}	Ra_{osc}	a^2_{stat}	a^2_{osc}	ϵ	Sg
129.810	5475.39	6.32	6.28	0.6	0.1
129.810	129.812	6.317	5.346	0.6	2.045
129.810	129.776	6.317	5.346	0.6	2.246
129.810	89.467	6.317	5.14	0.6	5
129.810	77.179	6.317	5.04	0.6	10
121.573	6771.24	6.22	6.14	0.7	0.1
121.573	121.586	6.217	5.271	0.7	2.584
121.573	121.561	6.217	5.270	0.7	2.585
121.573	92.563	6.217	5.13	0.7	5
121.573	78.611	6.217	5.04	0.7	10
115.231	8198.53	6.13	6.02	0.8	0.1
115.231	115.240	6.129	5.214	0.8	3.183
115.231	115.222	6.129	5.214	0.8	3.184
115.231	95.712	6.129	5.13	0.8	5
115.231	80.056	6.129	5.04	0.8	10
110.195	9757.51	6.05	5.93	0.9	0.1
110.195	110.206	6.051	5.171	0.9	3.841
110.195	110.193	6.051	5.171	0.9	3.842
110.195	98.913	6.051	5.13	0.9	5
110.195	81.514	6.051	5.04	0.9	10
106.097	11448.1	5.98	6.46	1	0.1
106.097	106.104	5.983	5.138	1	4.559
106.097	106.094	5.983	5.138	1	4.560
106.097	102.168	5.983	5.12	1	5
106.097	82.985	5.983	5.04	1	10

$$
\begin{aligned}
& w = v_3 = 0, \qquad \text{on} \qquad z = 0, d; \\
& T = T_L, \quad z = 0; \qquad T = T_U, \quad z = d; \\
& Q_1 = Q_2 = 0, \quad z = 0, d; \qquad Q_{3,3} = 0, \quad z = 0, d;
\end{aligned}
\tag{5.44}
$$

where $T_L > T_U$, with T_L, T_U both constants.

The steady solution of interest is the conduction solution

$$
\bar{v}_i \equiv 0, \qquad \bar{T} = -\beta z + T_L, \qquad \bar{\mathbf{Q}} = (0, 0, \kappa\beta),
\tag{5.45}
$$

where

$$
\beta = \frac{T_L - T_U}{d}
$$

is the temperature gradient.

To study instability we introduce perturbations to the steady solution $\bar{v}_i, \bar{T}, \bar{\mathbf{Q}}, \bar{p}$ as u_i, θ, q_i and π and these are non-dimensionalized with the scales

Table 5.3 Critical Rayleigh and wave numbers; $Pr = 0.1$, $\epsilon = 0.1$.

Ra_{stat}	Ra_{osc}	a^2_{stat}	a^2_{osc}	ζ	Sg
323.396	977.154	6.54	8.47	0.1	0.1
323.396	323.472	6.543	7.298	0.1	0.245
323.396	322.106	6.543	7.293	0.1	0.246
323.396	114.481	6.543	5.85	0.1	1
390.598	1122.83	6.19	7.67	0.2	0.1
390.598	391.435	6.188	6.527	0.2	0.288
390.598	390.383	6.188	6.524	0.2	0.289
390.598	197.493	6.188	5.59	0.2	1
457.317	1283.09	5.96	7.22	0.3	0.1
457.317	457.859	5.965	6.133	0.3	0.331
457.317	456.989	5.965	6.131	0.3	0.332
457.317	277.401	5.965	5.46	0.3	1
523.762	1449.23	5.81	6.94	0.4	0.1
523.762	523.777	5.810	5.891	0.4	0.374
523.762	523.029	5.810	5.890	0.4	0.375
523.762	356.055	5.810	5.39	0.4	1
590.037	1618.31	5.70	6.74	0.5	0.1
590.037	590.152	5.696	5.730	0.5	0.416
590.037	589.491	5.696	5.728	0.5	0.417
590.037	434.078	5.696	5.34	0.5	1

Table 5.4 Critical Rayleigh and wave numbers; $Pr = 0.1$, $\epsilon = 0.1$.

Ra_{stat}	Ra_{osc}	a^2_{stat}	a^2_{osc}	ζ	Sg
656.197	1789.09	5.61	6.60	0.6	0.1
656.197	656.304	5.609	5.613	0.6	0.458
656.197	655.709	5.609	5.612	0.6	0.459
656.197	511.738	5.609	5.30	0.6	1
722.278	1960.91	5.54	6.60	0.7	0.1
722.278	722.323	5.539	5.525	0.7	0.500
722.278	721.782	5.539	5.524	0.7	0.501
722.278	589.172	5.539	5.27	0.7	1
788.299	2133.44	5.48	6.40	0.8	0.1
788.299	788.756	5.483	5.457	0.8	0.541
788.299	788.259	5.483	5.456	0.8	0.542
788.299	666.454	5.483	5.25	0.8	1
854.277	2306.45	5.44	6.33	0.9	0.1
854.277	854.600	5.436	5.402	0.9	0.583
854.277	854.140	5.436	5.401	0.9	0.584
854.277	743.629	5.436	5.23	0.9	1
920.022	2479.82	5.40	6.27	1	0.1
920.022	920.412	5.397	5.357	1	0.625
920.022	919.984	5.397	5.357	1	0.626
920.022	820.726	5.397	5.22	1	1

$$
\begin{aligned}
&x_i = x_i^* d, \qquad u_i = u_i^* U, \qquad \theta = \theta^* T^\sharp, \qquad U = \frac{\kappa}{d},\\
&q_i = q_i^* Q^\sharp, \qquad \pi = \pi^* P, \qquad t = t^* \mathcal{T}, \qquad \hat{\tau} = \tau_f/\tau,\\
&\mathcal{T} = \frac{\rho}{\mu K}, \qquad Q^\sharp = \frac{\kappa T^\sharp}{d}, \qquad \xi_1 = \frac{\hat{\xi}_1}{d^2}, \qquad \xi_2 = \frac{\hat{\xi}_2}{d^2},
\end{aligned}
$$

and with the Darcy number, Prandtl number, parameter Sg, and Rayleigh number $Ra = R^2$,

$$Da = \frac{K}{d^2}, \qquad Pr = \frac{\mu}{\rho\kappa}, \qquad Sg = \frac{\tau\mu}{\rho d^2}, \qquad Ra = \frac{\alpha g d^2 \beta K \rho}{\mu\kappa}.$$

The non-dimensional perturbation equations are found to be

$$\begin{aligned}
&0 = -\pi_{,i} + R\theta k_i - u_i\,,\\
&u_{i,i} = 0,\\
&\frac{Pr}{MDa}\theta_{,t} + u_i\theta_{,i} = Rw - q_{i,i} + \zeta\Delta\theta,\\
&\frac{Sg}{Da}q_{i,t} + \frac{Sg}{Da}\hat{\tau}(u_j q_{i,j} - q_j u_{i,j}) = -q_i - \theta_{,i} + \frac{SgR}{Pr}\hat{\tau}u_{i,3} + \xi_1\Delta q_i + \xi_2 q_{j,ij}\,.
\end{aligned} \tag{5.46}$$

The solution to (5.46) is periodic in x, y and satisfies the boundary conditions

$$w = 0, \quad \theta = 0, \quad q_1 = q_2 = 0, \quad q_{3,3} = 0, \qquad z = 0, 1. \tag{5.47}$$

We now analyze the linear instability of the steady solution (5.45), and we let

$$P_1 = \frac{Pr}{MDa}, \qquad S_1 = \frac{Sg}{Da}.$$

To investigate instability we linearize (5.46) and remove the pressure from the equation for u_i, and take the divergence of $(5.46)_4$. Denoting $\Gamma = q_{i,i}$ and assuming a time dependence like $e^{\sigma t}$, one may show that this leads to the equations

$$\begin{aligned}
0 &= -R\Delta^*\theta + \Delta w,\\
P_1\sigma\theta &= Rw - \Gamma + \zeta\Delta\theta,\\
S_1\sigma\Gamma &= -\Gamma - \Delta\theta + \epsilon\Delta\Gamma.
\end{aligned} \tag{5.48}$$

Solving (5.48) subject to the boundary conditions $\theta = 0, w = 0, \Gamma = 0$ leads to the equation

$$\begin{aligned}
S_1P_1\Lambda\sigma^2 + \sigma\big[S_1\zeta\Lambda^2 + P_1\Lambda(1+\epsilon\Lambda)\big]&\\
+\zeta\Lambda^2(1+\epsilon\Lambda) + \Lambda^2 = R^2a^2(\sigma S_1 + 1 + \epsilon\Lambda)&.
\end{aligned} \tag{5.49}$$

With $\sigma = i\omega$, the real and imaginary parts of this equation yield

$$-\omega^2 S_1P_1\Lambda + \zeta\Lambda^2(1+\epsilon\Lambda) + \Lambda^2 = R^2a^2(1+\epsilon\Lambda), \tag{5.50}$$

and

$$S_1\zeta\Lambda^2 + P_1\Lambda(1+\epsilon\Lambda) = R^2a^2S_1\,. \tag{5.51}$$

Thus, the stationary convection boundary follows from

$$R^2_{stat} = \frac{\Lambda^2}{a^2(1+\epsilon\Lambda)} + \zeta\frac{\Lambda^2}{a^2}, \tag{5.52}$$

Table 5.5 Critical Rayleigh and wave numbers for stationary convection in a Darcy porous material; $Pr = 0.01$, $\epsilon = 0.01$.

Ra_{stat}	a^2_{stat}	ζ
36.6859	11.73	0.1
40.6598	11.50	0.2
44.6284	11.32	0.3
48.5931	11.18	0.4
52.5550	11.06	0.5
56.5147	10.97	0.6
60.4727	10.88	0.7
64.4294	10.81	0.8
68.3849	10.75	0.9
72.3396	10.70	1

while the oscillatory convection boundary is derived from

$$R^2_{osc} = \frac{P_1}{S_1}\frac{\Lambda}{a^2}(1+\epsilon\Lambda) + \zeta\frac{\Lambda^2}{a^2}\,. \tag{5.53}$$

The critical Rayleigh and wave numbers are found by minimizing (5.52) and (5.53) in a^2.

We may eliminate R^2a^2 from (5.50) and (5.51) to obtain

$$\omega^2 = \frac{\Lambda}{S_1P_1} - \frac{(1+\epsilon\Lambda)^2}{S_1^2}\,. \tag{5.54}$$

Since $\omega \in \mathbb{R}$, $\omega^2 > 0$, and so we find no oscillatory convection is possible if $2\epsilon \geq P_1S_1$. Certainly (5.54) is not valid for all a^2.

For many situations it is likely that stationary convection will be the dominant mechanism of convective motion. In tables 5.5 and 5.6 we display the Rayleigh and wave number instability thresholds as ϵ and ζ vary. Table 5.5 fixes $\epsilon = 0.01$ and displays the variation of R^2_{stat} and a^2_{stat} as ζ varies. Table 5.6 fixes $\zeta = 0.1$ and displays the variation of R^2_{stat} and a^2_{stat} as ϵ varies. We see that ϵ increasing lowers the critical Rayleigh number and so the Guyer–Krumhansl terms are destabilizing. However, when ζ is increased the critical Rayleigh number also increases and so the Mariano diffusion term does stabilize the layer.

Over the range of values displayed the wavenumber decreases as ζ is increased, meaning the convection cells become wider. The variation of a^2 is more marked as ϵ increases though, and a^2 rises more rapidly. This means that the convection cells become much narrower and this is associated with the fact that the layer is becoming more unstable.

Table 5.6 Critical Rayleigh and wave numbers for stationary convection in a Darcy porous material; $Pr = 0.01$, $\zeta = 0.1$.

Ra_{stat}	a^2_{stat}	ϵ
36.6859	11.73	0.01
31.6134	13.69	0.02
27.7305	15.63	0.03
24.7416	17.35	0.04
22.3383	18.75	0.05
20.4373	19.78	0.06
18.8921	20.48	0.07
17.6165	20.92	0.08
16.5478	21.17	0.09
15.6404	21.28	0.1

5.4.2 Effect of fluid inertia

The analogous instability problem to that studied in section 5.4.1 incorporating fluid inertia but without Mariano or Guyer–Krumhansl terms is described by Straughan (2011a, pages 238-241). The basic equations may be found there. We here add Guyer–Krumhansl terms to the evolution equation for the heat flux, and we add a Mariano extra flux term to the energy balance equation for the temperature field. Thus, the governing equations for thermal convection in this case are

$$\begin{aligned} &\frac{\partial v_i}{\partial t} = -\frac{1}{\rho}\frac{\partial p}{\partial x_i} + \alpha g k_i T - \frac{\mu}{\rho K} v_i\,, \\ &v_{i,i} = 0, \\ &\frac{1}{M}\frac{\partial T}{\partial t} + v_i \frac{\partial T}{\partial x_i} = -\frac{\partial Q_i}{\partial x_i} + \hat{\zeta}\Delta T, \\ &\tau\frac{\partial Q_i}{\partial t} + \tau_f\Big(v_j\frac{\partial Q_i}{\partial x_j} - Q_j\frac{\partial v_i}{\partial x_j}\Big) = -Q_i - \kappa\frac{\partial T}{\partial x_i} + \hat{\xi}_1\Delta Q_i + \hat{\xi}_2 Q_{j,ij}\,. \end{aligned} \tag{5.55}$$

The notation is as in section 5.4.1. The boundary conditions are again (5.44).

The stationary solution of interest now is (5.45). Perturbations are introduced and non-dimensionalized as in section 5.4.1. We now find the non-dimensional perturbation equations are

$$\begin{aligned} &u_{i,t} = -\pi_{,i} + R\theta k_i - u_i\,, \\ &u_{i,i} = 0, \\ &\frac{Pr}{MDa}\theta_{,t} + u_i\theta_{,i} = Rw - q_{i,i} + \zeta\Delta\theta, \\ &\frac{Sg}{Da}q_{i,t} + \frac{Sg}{Da}\hat{\tau}(u_j q_{i,j} - q_j u_{i,j}) = -q_i - \theta_{,i} + \frac{SgR}{Pr}\hat{\tau}u_{i,3} + \xi_1\Delta q_i + \xi_2 q_{j,ij}\,. \end{aligned} \tag{5.56}$$

The solution to (5.56) satisfies a plane tiling periodic pattern in the x, y-plane, and additionally satisfies the boundary conditions

$$w = 0, \quad \theta = 0, \quad q_1 = q_2 = 0, \quad q_{3,3} = 0, \qquad z = 0, 1. \tag{5.57}$$

To analyse instability we proceed in a manner analogous to section 5.4.1. Instead of (5.48) we now obtain

$$\begin{aligned} -\sigma\Delta w &= -R\Delta^*\theta + \Delta w, \\ P_1\sigma\theta &= Rw - \Gamma + \zeta\Delta\theta, \\ S_1\sigma\Gamma &= -\Gamma - \Delta\theta + \epsilon\Delta\Gamma. \end{aligned} \tag{5.58}$$

Solution of these equations with the relevant boundary conditions results in the cubic equation

$$\begin{aligned} S_1P_1\Lambda\sigma^3 &+ \sigma^2\big[P_1\Lambda(1+\epsilon\Lambda) + S_1(P_1\Lambda + \zeta\Lambda^2)\big] \\ &+ \sigma\big[(P_1\Lambda + \zeta\Lambda^2)(1+\epsilon\Lambda) + S_1\zeta\Lambda^2\big] + \zeta\Lambda^2(1+\epsilon\Lambda) \\ &+ (\sigma+1)\Lambda^2 = R^2a^2(\sigma S_1 + 1 + \epsilon\Lambda). \end{aligned} \tag{5.59}$$

The stationary convection equation which follows from this is

$$R^2_{stat} = \frac{\Lambda^2}{a^2(1+\epsilon\Lambda)} + \zeta\frac{\Lambda^2}{a^2}\,. \tag{5.60}$$

To find the oscillatory convection boundary we put $\sigma = i\omega$ in (5.59) and take the real and imaginary parts of the result to obtain the equations

$$-\omega^2\big[P_1\Lambda(1+\epsilon\Lambda) + S_1(P_1\Lambda + \zeta\Lambda^2)\big] + \zeta\Lambda^2(1+\epsilon\Lambda) + \Lambda^2 = R^2a^2(1+\epsilon\Lambda), \tag{5.61}$$

and

$$-\omega^2P_1S_1\Lambda + (P_1\Lambda + \zeta\Lambda^2)(1+\epsilon\Lambda) + S_1\zeta\Lambda^2 + \Lambda^2 = R^2a^2S_1\,. \tag{5.62}$$

These equations lead to the oscillatory convection equation

$$\begin{aligned} R^2_{osc} =& \zeta\frac{\Lambda^2}{a^2} + \frac{\Lambda}{a^2}(1+\epsilon\Lambda)^2\frac{P_1}{S_1^2} + \frac{\Lambda}{a^2}\frac{(P_1+\zeta\Lambda)(1+\epsilon\Lambda)}{S_1^2} \\ &+ \frac{\Lambda^2}{a^2}\frac{P_1(1+\epsilon\Lambda)}{S_1^2(P_1+\zeta\Lambda)} + \frac{\Lambda^3}{a^2}\frac{\zeta}{S_1(P_1+\zeta\Lambda)}\,. \end{aligned} \tag{5.63}$$

The critical Rayleigh and wave numbers are found from (5.60) and (5.63) by minimizing in a^2.

If we eliminate R^2a^2 from (5.61) and (5.62) then one may obtain

$$\omega^2 = \frac{S_1(P_1\Lambda + \zeta\Lambda^2)}{(1+\epsilon\Lambda)} + \frac{\Lambda^2}{1+\epsilon\Lambda} - \frac{\Lambda^2}{S_1} - \frac{(P_1\Lambda + \zeta\Lambda^2)(1+\epsilon\Lambda)}{S_1}. \tag{5.64}$$

For oscillatory convection $\omega^2 > 0$ and so (5.64) leads to a constraint which must be checked in conjunction with the minimization of (5.63).

We do not provide detailed numerical results, but we point out that (5.60) is the same equation as one obtains in the zero inertia case analysed in section 5.4.1. Thus, the results in tables 5.5 and 5.6 apply also to the problem of this section. Likewise, the comments at the end of section 5.4.1 apply here.

Chapter 6
Kelvin–Voigt fluids

6.1 Cattaneo convection in a Navier–Stokes–Voigt fluid

The equations for convection in a Navier–Stokes–Voigt fluid were already introduced in section 2.6, they are given by (2.47) when Fourier's law is employed for the heat flux. In fact, the equations for thermal convection in a Navier–Stokes–Voigt fluid with a Cattaneo–Christov theory for the heat flux were introduced in section 4.4.2. Since the object of that section was to obtain a nonlinear energy stability theory in three space dimensions a Mariano effect was also incorporated into the energy balance equation as well as a Laplacian of the heat flux in the Cattaneo equation.

Our object in this section is to analyse the linear instability of the steady conduction solution for a Navier–Stokes–Voigt fluid with CCJ theory or with Cattaneo–Christov theory, but without a Mariano effect and no Laplacian of the heat flux. The steady conduction solution is the usual one with $\bar{v}_i \equiv 0$, the temperature is linear in z, and the third component of the heat flux is constant. The nonlinear non-dimensional perturbation equations then follow from (4.49) by choosing $\zeta = 0$ and $\epsilon = 0$.

Thus, the Cattaneo–Christov theory yields the relevant perturbation equations as

$$
\begin{aligned}
&u_{i,t} + u_j u_{i,j} - \lambda \Delta u_{i,t} = -\pi_{,i} + \Delta u_i + R\theta k_i\,,\\
&u_{i,i} = 0,\\
&Pr(\theta_{,t} + u_i \theta_{,i}) = Rw - q_{i,i}\,,\\
&Sg(q_{i,t} + u_j q_{i,j} - u_{i,j} q_j) = \frac{RSg}{Pr}\, u_{i,3} - q_i - \theta_{,i}\,.
\end{aligned}
\tag{6.1}
$$

The analogous equations for CCJ theory do not have the RSg/Pr term nor the $-Sgu_{i,j}q_j$ term in $(6.1)_4$. The boundary conditions are

$$
u_i = 0, \quad \theta = 0, \qquad z = 0, 1, \tag{6.2}
$$

with the solution being periodic in x, y. It is worth observing that system (6.1) contains a skew - symmetric part to the linear operator which leads to the presence of oscillatory instability, as we show below.

B. Straughan, *Thermal Convection with a Cattaneo Flux Law*, Advances in Mechanics and Mathematics 53, https://doi.org/10.1007/978-3-031-80885-2_6

We are here interested in a linear instability analysis and so we discard the nonlinear terms in (6.1) and assume a time dependence like $u_i = e^{\sigma t} u_i(\mathbf{x})$, with similar forms for θ, π and q_i. This leads to the system of linear equations

$$
\begin{aligned}
&\sigma(1 - \lambda\Delta)u_i = -\pi_{,i} + \Delta u_i + R\theta k_i\,,\\
&u_{i,i} = 0,\\
&Pr\sigma\theta = Rw - q_{i,i}\,,\\
&\sigma Sg q_i = \frac{RSg}{Pr} u_{i,3} - q_i - \theta_{,i}\,.
\end{aligned}
\tag{6.3}
$$

The pressure is removed using the fact that the velocity field is solenoidal and we take the divergence of $(6.3)_4$ to reduce the equations to the system

$$
\begin{aligned}
&-\sigma\Delta w + \lambda\sigma\Delta^2 w = -\Delta^2 w - R\Delta^*\theta,\\
&Pr\sigma\theta = Rw - \Gamma,\\
&\sigma Sg\Gamma = -\Gamma - \Delta\theta.
\end{aligned}
\tag{6.4}
$$

For two stress free surfaces we may represent w, θ and Γ as a series in $\sin n\pi z$ and this leads to the equation

$$
\begin{aligned}
PrSg(\Lambda_n + \lambda\Lambda_n^2)\sigma^3 &+ \sigma^2\big[Pr(\Lambda_n + \lambda\Lambda_n^2) + PrSg\Lambda_n^2\big]\\
&+ \sigma(Pr\Lambda_n^2 + \Lambda_n^2 + \lambda\Lambda_n^3) + \Lambda_n^3 = R^2a^2(\sigma Sg + 1),
\end{aligned}
\tag{6.5}
$$

where $\Lambda_n = n^2\pi^2 + a^2$, a being the wavenumber.

The stationary convection threshold follows from (6.5) as

$$
R^2_{stat} = \frac{\Lambda_n^3}{a^2}\,,
$$

and the minimum in n yields

$$
R^2_{stat} = \frac{\Lambda^3}{a^2}\,,
$$

with $\Lambda = \pi^2 + a^2$, and then the minimum in a^2 yields

$$
a^2_{stat} = \frac{\pi^2}{2}\,, \qquad R^2_{stat} = \frac{27\pi^4}{4}\,.
$$

For oscillatory convection put $\sigma = i\omega$ in (6.5) and then take the real and imaginary parts of the resulting equation. One solves these for ω^2 and then one determines the oscillatory convection Rayleigh number as

$$
R^2_{osc} = \left(\frac{Pr + 1 + \lambda\Lambda_n}{Sg^2}\right)\left(\frac{\Lambda_n}{a^2} + \lambda\frac{\Lambda_n^2}{a^2}\right) + \frac{Pr\Lambda_n^2}{Sga^2}\,.
$$

The minimum of this in n is given with $n = 1$ and then we may write

$$\begin{aligned} R^2_{osc} &= \frac{(Pr+1)}{S_g^2}\frac{\Lambda}{a^2} + \frac{Pr}{Sg}\frac{\Lambda^2}{a^2} + \frac{\lambda(Pr+2)}{Sg^2}\frac{\Lambda^2}{a^2} + \frac{\lambda^2}{S_g^2}\frac{\Lambda^3}{a^2}, \\ &= R^2_{osc}NS + \frac{\lambda(Pr+2)}{Sg^2}\frac{\Lambda^2}{a^2} + \frac{\lambda^2}{Sg^2}\frac{\Lambda^3}{a^2}, \end{aligned} \tag{6.6}$$

where $R^2_{osc}NS$ is the equivalent expression one finds for Cattaneo convection with a Navier–Stokes fluid, cf. Straughan (2010b). If we denote the Navier–Stokes–Voigt expression by $R^2_{osc}KV$ then, $R^2_{osc}KV \geq R^2_{osc}NS$, and also we see that when $\lambda^2/Sg^2 \geq 1$ then $R^2_{osc}KV \geq R^2_{stat}$ and so if $\lambda^2 > Sg^2$ we do not witness oscillatory convection.

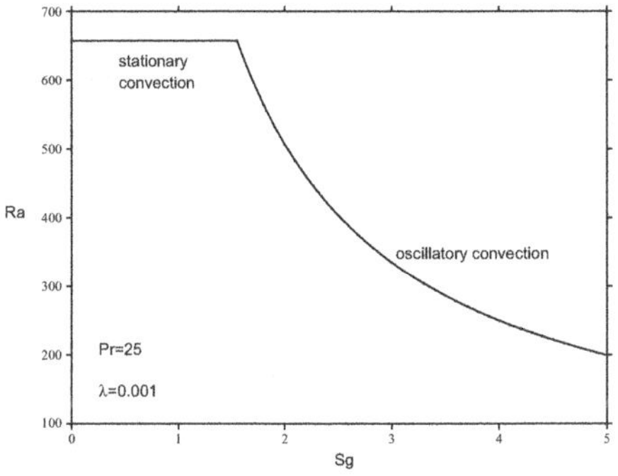

Fig. 6.1 Navier–Stokes–Voigt fluid, Ra vs. Sg with $Pr = 25$, $\lambda = 0.001$.

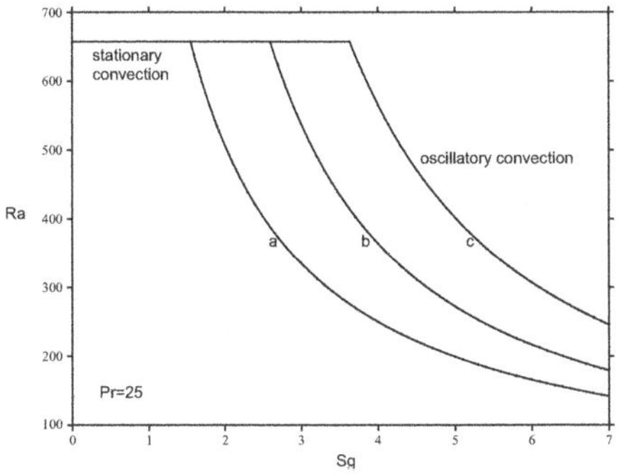

Fig. 6.2 Navier–Stokes–Voigt fluid, Ra vs. Sg with $Pr = 25$. Curve a is for $\lambda = 0$, curve b is for $\lambda = 1$, curve c is for $\lambda = 2$.

The behaviour of the critical values for R^2_{osc} and a^2_{osc} is displayed in figures 6.1 and 6.2 and table 6.1. Figure 6.1 with $\lambda = 10^{-3}$ is given primarily for comparison with work in a Kelvin–Voigt fluid or order one in section 6.3, although the behaviour is very similar to that when $\lambda = 0$ shown in figure 6.2. For each value of the Kelvin–Voigt coefficient λ the instability picture has a similar shape in that for Sg small enough stationary convection is found and then depending on λ a threshold value of Sg is reached such that oscillatory convection becomes the dominant mechanism of convective motion. The oscillatory convection threshold increases with increasing λ as shown in table 6.1. In figure 6.1 the conduction solution is unstable for Ra values

Table 6.1 Navier–Stokes–Voigt fluid. Critical oscillatory wave numbers, λ values, and Sg values at the transition from stationary to oscillatory convection. The critical Rayleigh number at the transition is $Ra = 657.511$, and the critical stationary wavemumber squared is $\pi^2/2 \approx 4.9348$.

a_{osc}^2	λ	Sg
10.199	0	1.552
10.199	0.001	1.553
8.565	1	2.597
7.505	2	3.634
6.292	5	6.685
5.695	10	11.71

above the curve. The oscillatory convection wavenumber decreases as λ is increased as is seen in table 6.1. This means the convection cells become wider as λ increases.

6.2 Kelvin–Voigt fluid of order N

Thermal convection in a Kelvin–Voigt fluid of variable order was analysed in Straughan (2022b) for the situation where Fourier's law holds for the heat flux. He observes that interesting models for Kelvin–Voigt fluids have been presented by Oskolkov (1988, 1995), Oskolkov and Shadiev (1994). Generalizations of the Kelvin–Voigt models to incorporate thermal effects are given by Sukacheva and Matveeva (2010), Matveeva (2013), Sukacheva and Kondyukov (2014). The latter pieces of work mainly concentrate on solvability issues and on establishing properties of solutions such as existence of a suitable solution and uniqueness questions.

Straughan (2022b) presented analysis of thermal convection of a layer of Kelvin–Voigt fluid heated from below employing the models of Sukacheva and Matveeva (2010), Sukacheva and Kondyukov (2014). He presented a computational procedure for calculating the neutral curves for instability associated with thermal convection. This is a non-trivial process for Kelvin–Voigt fluids of variable order since it transpires that to calculate the critical Rayleigh number thresholds of instability one has to first calculate the eigenvalues of the variable order theory and then employ these in the minimization process over all wave numbers.

We now describe the equations for thermal convection in a Kelvin–Voigt fluid of order $0, 1, 2, \ldots, L$.

Let $v(\mathbf{x}, t)$, $T(\mathbf{x}, t)$, $p(\mathbf{x}, t)$ be the velocity, temperature and pressure at position $\mathbf{x}$ and time t of a body of fluid. The Kelvin–Voigt equations of non-isothermal fluid flow of variable order may be written, see Sukacheva and Matveeva (2010),

$$
\begin{aligned}
(1-\hat{\lambda}\Delta)v_{i,t}+v_jv_{i,j} &= -\frac{1}{\rho_0}p_{,i}+\nu\Delta v_i+\alpha Tgk_i+\sum_{\alpha=1}^{L}\beta_\alpha\Delta W_i^\alpha,\\
v_{i,i} &= 0,\\
T_{,t}+v_iT_{,i} &= k\Delta T,\\
W_{i,t}^\alpha+\hat{\gamma}^\alpha W_i^\alpha &= v_i, \qquad \alpha=1,2,\dots,L.
\end{aligned}
\tag{6.7}
$$

Here $\hat{\lambda},\nu,\rho_0,\alpha,g$ and k are the Kelvin–Voigt coefficient, the kinematic viscosity, a reference density, the coefficient of thermal expansion of the fluid, gravity, and thermal diffusivity. The symbol Δ denotes the Laplacian, $k_i\equiv\mathbf{k}=(0,0,1)$, (observe that k_i and k are different),

The terms W_i^α represent the viscoelastic fading memory effect, and $\hat{\gamma}^\alpha>0$, are constants, $\alpha=1,\dots,L$. In deriving equations (6.7) a Boussinesq approximation has been employed whereby the density in the buoyancy force is written as $\rho=\rho_0[1-\alpha(T-T_0)]$ for a reference temperature T_0 and this gives rise to the αgTk_i term with the constants appearing as a hydrostatic pressure, cf. Straughan (2004), pp. 47-49, Straughan (2015a), pp. 16-21.

As Oskolkov (1988, 1995) points out, the W_i^α terms are equivalent to a stress relation of form

$$
\sigma_{ij}=\kappa_3\frac{\partial d_{ij}}{\partial t}+\nu d_{ij}+\sum_{\beta=1}^{L}\xi_\beta\int_{-\infty}^{t}\exp\{-\hat{\gamma}_\beta(t-s)\}\,d_{ij}\,ds,
$$

where $\kappa_3>0$ and $\xi_\beta>0$ are constants, d_{ij} is the symmetric part of the velocity gradient, and σ_{ij} is the Cauchy extra stress tensor. This representation, Oskolkov (1988, 1995), also arises from a stress representation of form

$$
\sum_{\alpha=0}^{L}\lambda_\alpha\frac{\partial^\alpha}{\partial t^\alpha}\sigma_{ij}=\sum_{\beta=0}^{L+1}\kappa_\beta\frac{\partial^\beta}{\partial t^\beta}d_{ij},
$$

for suitable coefficients λ_α and κ_β. One has to be careful with interpreting the stress relation in a fully nonlinear theory as is pointed out by Damázio et al. (2016), Straughan (2022a). We here analyse only linear instability.

If $L=0$ then equations (6.7) are sometimes called the Navier–Stokes–Voigt equations, cf. Layton and Rebholz (2013), or sometimes the Kelvin–Voigt equations of order zero, or the Oskolkov equations, see Oskolkov (1995). For $L=1,2,\dots$ they are known as the Kelvin–Voigt equations of order L.

We suppose the fluid occupies the horizontal layer $0<z<d$ with gravity acting downward. Then equations (6.7) are defined on the spatial region $\mathbb{R}^2\times\{z\in(0,d)\}$.

The boundary conditions on the planes $z=0$ and $z=d$ are given by

$$
\begin{aligned}
&v_i=0,\quad z=0,d;\qquad T=T_L,\quad z=0,\qquad T=T_U,\quad z=d;\\
&W_i^\alpha=0,\quad z=0,d,\quad \alpha=1,\dots,L,
\end{aligned}
\tag{6.8}
$$

with constant values T_L °K, T_U °K, where °K denotes degrees Kelvin, and $T_L > T_U > 0$.

The steady solution analysed by Straughan (2022b) is given by

$$\bar{v}_i \equiv 0, \quad \bar{W}_i^\alpha \equiv 0, \quad \bar{T} = -\beta z + T_L, \tag{6.9}$$

where the temperature gradient, β, is given by

$$\beta = \frac{T_L - T_U}{d},$$

and the steady pressure $\bar{p}$ is a quadratic in z which follows from equation $(6.7)_1$.

To investigate stability of the solution (6.9) Straughan (2022b) introduced perturbations $(u_i, \theta, \pi, w_i^\alpha)$ by

$$v_i = \bar{v}_i + u_i, \quad T = \bar{T} + \theta, \quad p = \bar{p} + \pi, \quad W_i^\alpha = \bar{W}_i^\alpha + w_i^\alpha .$$

The equations for the perturbations are then derived and non-dimensionalized with the scalings

$$x_i = x_i^* d, \qquad t = t^* \mathcal{T}, \qquad U = \frac{\nu}{d}, \qquad \mathcal{T} = \frac{d^2}{\nu},$$

$$P = \frac{\rho_0 \nu U}{d}, \qquad W = \frac{\nu \mathcal{T}}{d}, \qquad T^\sharp = U\sqrt{\frac{\beta\nu}{kg\alpha}},$$

where $\mathcal{T}, U, W$ and $T^\sharp$ are the scales for time, velocity, the vectors W_i^α, and temperature, respectively. Define the Rayleigh number, $Ra = R^2$, and the Prandtl number Pr, by

$$Ra = \frac{\alpha g \beta d^4}{\nu k}, \qquad Pr = \frac{\nu}{k},$$

and rescale the coefficients $\hat{\gamma}_\alpha$, $\hat{\lambda}$, β_α by the non-dimensional ones

$$\gamma^\alpha = \hat{\gamma}^\alpha \mathcal{T}, \qquad \lambda = \frac{\hat{\lambda}}{\mathcal{T}\nu}, \qquad \epsilon_\alpha = \frac{\beta_\alpha d}{\nu U}.$$

If one drops the *s and treats x_i and t as the non-dimensional variables, the non-dimensional form of the perturbation equations to (6.7) as derived by Straughan (2022b) are

$$\begin{aligned}
&u_{i,t} - \lambda \Delta u_{i,t} + u_j u_{i,j} = -\pi_{,i} + \Delta u_i + R k_i \theta + \sum_{\alpha=1}^{L} \epsilon_\alpha \Delta w_i^\alpha, \\
&u_{i,i} = 0, \\
&Pr(\theta_{,t} + u_i \theta_{,i}) = Rw + \Delta\theta, \\
&w_{i,t}^\alpha + \gamma^\alpha w_i^\alpha = u_i, \qquad \alpha = 1, 2, \ldots, L,
\end{aligned} \tag{6.10}$$

where $w = u_3$.

Equations (6.10) are defined on the domain $\mathbb{R}^2 \times \{z \in (0,1)\} \times \{t > 0\}$ and the boundary conditions are

$$u_i = 0, \quad w_i^\alpha = 0, \quad \theta = 0, \qquad z = 0, 1, \tag{6.11}$$

together with the fact that u_i, w_i^α, θ and π satisfy a plane tiling periodicity in the x, y plane.

The object of this section is to present equations analogous to those of (6.7), but now where a Cattaneo equation is employed for the heat flux. We restrict attention to a CCJ theory since the notation is already complex, but it is straightforward to employ other Cattaneo theories.The relevant equations for a thermal convection in a Kelvin–Voigt fluid of order L which employs CCJ theory with a Cattaneo heat flux are

$$\begin{aligned}
&(1-\hat{\lambda}\Delta)v_{i,t} + v_j v_{i,j} = -\frac{1}{\rho_0}p_{,i} + \nu\Delta v_i + \alpha T g k_i + \sum_{\alpha=1}^{L}\beta_\alpha \Delta W_i^\alpha, \\
&v_{i,i} = 0, \\
&T_{,t} + v_i T_{,i} = -Q_{i,i}, \\
&\tau(Q_{i,t} + v_j Q_{i,j}) = -Q_i - kT_{,i}, \\
&W_{i,t}^\alpha + \hat{\gamma}^\alpha W_i^\alpha = v_i, \qquad \alpha = 1, 2, \dots, L,
\end{aligned} \tag{6.12}$$

where τ is the relaxation time in the Cattaneo equation.

The boundary conditions are the same as (6.8) and the steady solution is the same as (6.9) with the steady heat flux given by $\bar{\mathbf{Q}} = (0, 0, k\beta)$.

To analyse instability perturbation equations are developed as above in the Fourier case and similarly non-dimensionalized to obtain the perturbation equations in the Cattaneo case as

$$\begin{aligned}
&u_{i,t} - \lambda\Delta u_{i,t} + u_j u_{i,j} = -\pi_{,i} + \Delta u_i + Rk_i\theta + \sum_{\alpha=1}^{L}\epsilon_\alpha \Delta w_i^\alpha, \\
&u_{i,i} = 0, \\
&Pr(\theta_{,t} + u_i\theta_{,i}) = Rw - q_{i,i}, \\
&Sg(q_{i,t} + u_j q_{i,j}) = -q_i - \theta_{,i}, \\
&w_{i,t}^\alpha + \gamma^\alpha w_i^\alpha = u_i, \qquad \alpha = 1, 2, \dots, L.
\end{aligned} \tag{6.13}$$

We do not proceed further with the general case, but now analyse in detail the case where $L = 1$, that is, thermal convection in a Kelvin–Voigt fluid of order one.

6.3 Kelvin–Voigt fluid of order 1

To analyse thermal convection in a Kelvin–Voigt fluid of order one with a Cattaneo law for the heat flux we take $L = 1$ in (6.12), and then the relevant governing equations are

$$
\begin{aligned}
&(1-\hat{\lambda}\Delta)v_{i,t} + v_j v_{i,j} = -\frac{1}{\rho_0}p_{,i} + \nu\Delta v_i + \alpha T g k_i + \beta_1 \Delta W_i^1,\\
&v_{i,i} = 0,\\
&T_{,t} + v_i T_{,i} = -Q_{i,i},\\
&\tau(Q_{i,t} + v_j Q_{i,j}) = -Q_i - kT_{,i},\\
&W_{i,t}^1 + \hat{\gamma}^1 W_i^1 = v_i\,.
\end{aligned}
\tag{6.14}
$$

The boundary conditions and the steady state are as stated in section 6.2 and so the perturbation equations have form

$$
\begin{aligned}
&u_{i,t} - \lambda\Delta u_{i,t} + u_j u_{i,j} = -\pi_{,i} + \Delta u_i + Rk_i\theta + \underline{\underline{\beta\Delta w_i}},\\
&u_{i,i} = 0,\\
&Pr(\theta_{,t} + u_i\theta_{,i}) = Rw \underline{-q_{i,i}},\\
&Sg(q_{i,t} + u_j q_{i,j}) = -q_i \underline{-\theta_{,i}},\\
&w_{i,t} + \gamma w_i = \underline{\underline{u_i}},
\end{aligned}
\tag{6.15}
$$

where w_i replaces w_i^1, β, γ are non-dimensional forms of ϵ_1 and γ^1.

We observe that the linear operator in (6.15) contains two sets of skew-symmetric terms. These are $-q_{i,i}$ and $-\theta_{,i}$ shown underlined once, but also $\beta\Delta w_i$ and u_i shown by double underline. To see the latter pair yield a skew-symmetric part take $-\Delta$ of $(6.15)_5$. We expect the presence of two skew-symmetric parts to lead to an interesting instability surface.

The boundary conditions are

$$
u_i = 0, \quad \theta = 0, \quad w_i = 0, \qquad z = 0, 1,
$$

with $u_i, \pi, \theta, w_i, q_i$ being periodic in x, y. We develop a linear instability analysis and so we discard the nonlinear terms in (6.15) and we assume $u_i = e^{\sigma t}u_i(\mathbf{x})$, with a similar representation for π, θ, q_i and w_i. Then equation $(6.15)_5$ reduces to

$$
(\sigma + \gamma)w_i = u_i\,,
$$

and we employ this in $(6.15)_1$, put $\Gamma = q_{i,i}$, and take the divergence of equation $(6.15)_4$. This yields the system of equations

$$\begin{aligned}\sigma(1-\lambda\Delta)u_i &= -\pi_{,i}+\Delta u_i+R\theta k_i+\frac{\beta}{\sigma+\gamma}\Delta u_i\,,\\ Pr\sigma\theta &= Rw-\Gamma,\\ Sg\sigma\Gamma &= -\Gamma-\Delta\theta.\end{aligned}\tag{6.16}$$

We now remove the pressure in $(6.16)_1$ to further reduce the system to

$$\begin{aligned}-\sigma(1-\lambda\Delta)\Delta w &= -\Delta^2 w-R\Delta^*\theta-\frac{\beta}{\sigma+\gamma}\Delta^2 w\,,\\ Pr\sigma\theta &= Rw-\Gamma,\\ Sg\sigma\Gamma &= -\Gamma-\Delta\theta.\end{aligned}\tag{6.17}$$

For two stress free surfaces we write w, θ and Γ as sin series and then after some calculation one may reduce the system to the equation

$$\begin{aligned}&\sigma Pr(Sg\sigma+1)\Big[\sigma(\Lambda+\lambda\Lambda^2)+\Lambda^2+\frac{\beta\Lambda^2}{\sigma+\gamma}\Big]\\ &\qquad+\Lambda\Big[\sigma(\Lambda+\lambda\Lambda^2)+\Lambda^2+\frac{\beta\Lambda^2}{\sigma+\gamma}\Big]=R^2a^2(Sg\sigma+1).\end{aligned}\tag{6.18}$$

The stationary convection boundary follows from (6.18) by taking $\sigma = 0$, and so

$$R^2_{stat}=\frac{\Lambda^3}{a^2}\Big(1+\frac{\beta}{\gamma}\Big),$$

and minimizing in a^2,

$$R^2_{stat}=\frac{27\pi^4}{4}\Big(1+\frac{\beta}{\gamma}\Big),\tag{6.19}$$

and

$$a^2_{stat}=\frac{\pi^2}{2}\,.\tag{6.20}$$

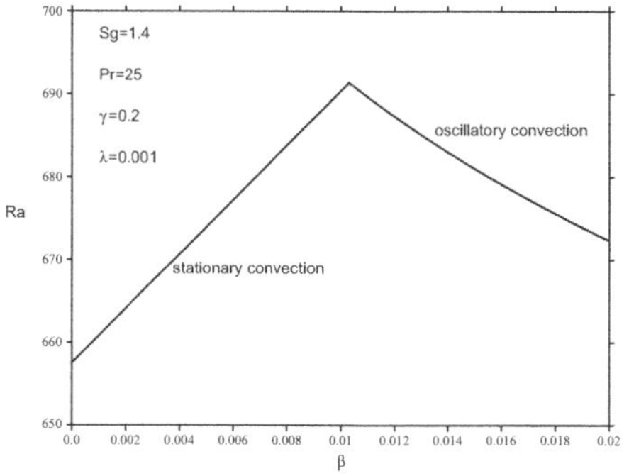

Fig. 6.3 Ra vs. β with $Pr = 25$, $\zeta = 0.2$, $Sg = 1.4$, $\lambda = 0.001$. The straight line emanating from $Ra = 657.511$, $\beta = 0$ the stationary convection branch which has a transition at $Ra = 691.37$, $\beta = 0.0103$, and the curve continues with the oscillatory convection branch which decreases in Ra from the transition point. Detail of Ra near the transition from stationary to oscillatory convection.

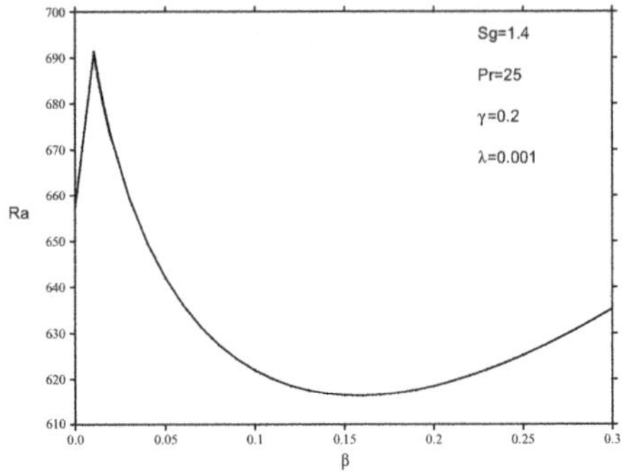

Fig. 6.4 Ra vs. β with $Pr = 25$, $\zeta = 0.2$, $Sg = 1.4$, $\lambda = 0.001$.. Further detail of Ra on the oscillatory convection curve.

One now takes the real and imaginary parts of (6.18) and after some rearrangement one may show

$$
\begin{aligned}
R^2a^2 =& -\omega^2\Big[Pr\Big(\Lambda + \lambda\Lambda^2 - \frac{\beta\Lambda^2}{\gamma^2+\omega^2}\Big) - \frac{Sg}{(1+\omega^2 Sg^2)}\Big(\Lambda^2 + \lambda\Lambda^3 - \frac{\beta\Lambda^3}{\gamma^2+\omega^2}\Big)\Big] \\
&+ \frac{1}{(1+\omega^2 Sg^2)}\Big(\Lambda^3 + \frac{\beta\Lambda^3\gamma}{\gamma^2+\omega^2}\Big) \\
&+ i\omega\Big\{Pr\Big(\Lambda^2 + \frac{\beta\Lambda^2\gamma}{\gamma^2+\omega^2}\Big) + \frac{1}{(1+\omega^2 Sg^2)}\Big(\Lambda^2 + \lambda\Lambda^3 - \frac{\beta\Lambda^3}{\gamma^2+\omega^2}\Big) \\
&- \frac{Sg}{(1+\omega^2 Sg^2)}\Big(\Lambda^3 + \frac{\beta\Lambda^3\gamma}{\gamma^2+\omega^2}\Big)\Big\}.
\end{aligned}
\tag{6.21}
$$

Since R^2 is real it follows that the coefficient of $i\omega$ is zero and this yields a quadratic in ω^2 of form

$$X\omega^4 + Y\omega^2 + Z = 0,$$

where

$$
\begin{aligned}
X &= Pr\Lambda^2 Sg^2, \\
Y &= Pr\Lambda^2(Sg^2\gamma^2 + 1) + Pr\beta\gamma\Lambda^2 Sg^2 + \Lambda^2 + \Lambda^3(\lambda - Sg), \\
Z &= Pr\Lambda^2(\gamma^2 + \beta\gamma) + \gamma^2\big[\Lambda^2 + \Lambda^3(\lambda - Sg)\big] - (1 + Sg\gamma)\beta\Lambda^3.
\end{aligned}
$$

The solutions are given by

$$\omega^2 = \frac{-Y \pm \sqrt{Y^2 - 4XZ}}{2X}.$$

We solve this numerically and find only $\omega_1^2 > 0$ in certain parameter ranges.

The critical Rayleigh numbers are then found by using the appropriate values of $\omega_1^2 > 0$ in (6.21), recalling the $i\omega$ term is not present, and by then minimizing R^2 in a^2. This yields the oscillatory convection Rayleigh number as $R^2_{osc}(Sg, \beta, \gamma, \lambda, Pr)$.

Table 6.2 Critical Rayleigh numbers and oscillatory wave numbers, the stationary wavenumber squared is $\pi^2/2 \approx 4.9348$; $Pr = 25$, $\gamma = 0.2$, $Sg = 0.1$, $\lambda = 0.001$. Where there is no entry signifies $\omega^2 < 0$.

Ra_{stat}	Ra_{osc}	a^2_{osc}	β
690.387	-	-	0.01
723.263	-	-	0.02
756.138	-	-	0.03
789.014	-	-	0.04
821.889	-	-	0.05
854.765	-	-	0.06
887.640	-	-	0.07
920.516	-	-	0.08
953.391	959.486	5.915	0.09
961.939	961.968	5.922	0.0926
986.267	969.028	5.940	0.1
1019.14	978.559	5.965	0.11
1052.02	988.078	5.990	0.12
1084.89	997.587	6.015	0.13
1117.77	1007.09	6.039	0.14
1150.64	1016.57	6.062	0.15
1183.52	1026.05	6.086	0.16
1216.40	1035.52	6.109	0.17
1249.27	1044.98	6.132	0.18
1282.15	1054.43	6.155	0.19
1315.02	1063.43	6.177	0.2

The instability surface depends on five parameters and consists of R^2_{stat} or R^2_{osc} depending on which is smallest over the appropriate parameter range. We may clearly create many numerical results. We select $Pr = 25, \gamma = 0.2, \lambda = 10^{-3}$, to allow comparison with results in section 6.1 and with results given in Straughan (2022b) for the Kelvin–Voigt order one Fourier case. The instability surface is then a two-dimensional surface in R^2, Sg, β space. When $\beta = 0$ the curve is given by figure 6.1. When $Sg = 0$ the instability boundary is shown in figure one of Straughan (2022b, page 196), and consists of a straight line for stationary convection increasing from $Ra = 27\pi^4/4$ at $Sg = 0$ to a transition point where oscillatory convection takes over with another straight line of lesser slope. The instability surface in the case $\beta > 0, Sg > 0$ is a complicated one. Figure 6.3 shows the behaviour when $Sg = 1.4$ and β varies, for small β. Figure 6.4 is a continuation of this curve for larger values of β. One sees that this is not a simple shape; once oscillatory convection commences the curve decreases to a minimum at $\beta = 0.16$ with $R^2_{osc} = 616.357$ and thereafter increases. We did not witness this behaviour for $S \leq 0.8$, but the minimum is also seen for $Sg \geq 0.9$. Actual values of the oscillatory surface are given in tables 6.2 - 6.7. The minimum on the Ra versus β curve is not fixed in β, for example for $Sg = 0.9$, $Ra^{osc}_{min} = 779.209$, when $\beta = 0.04$, whereas for $Sg = 1$, one finds $Ra^{osc}_{min} = 750.661$, when $\beta = 0.11$.

Table 6.3 Critical Rayleigh numbers and oscillatory wave numbers, the stationary wavenumber squared is $\pi^2/2 \approx 4.9348$; $Pr = 25, \gamma = 0.2, Sg = 0.7, \lambda = 0.001$. Where there is no entry signifies $\omega^2 < 0$.

Ra_{stat}	Ra_{osc}	a^2_{osc}	β
690.387	-	-	0.01
723.263	-	-	0.02
756.138	-	-	0.03
789.014	-	-	0.04
818.930	818.950	6.025	0.0491
821.889	819.239	6.030	0.05
854.765	822.482	6.087	0.06
887.640	825.774	6.143	0.07
920.516	829.115	6.198	0.08
953.391	832.504	6.253	0.09
986.267	835.939	6.308	0.1
1019.14	839.419	6.361	0.11
1052.02	842.943	6.414	0.12
1084.89	846.511	6.467	0.13
1117.77	850.121	6.519	0.14
1150.64	853.772	6.571	0.15
1183.52	857.464	6.622	0.16
1216.40	861.195	6.672	0.17
1249.27	864.965	6.722	0.18
1282.15	868.774	6.771	0.19
1315.02	872.619	6.820	0.2

Table 6.4 Critical Rayleigh numbers and oscillatory wave numbers, the stationary wavenumber squared is $\pi^2/2 \approx 4.9348$; $Pr = 25, \gamma = 0.2, Sg = 1, \lambda = 0.001$. Where there is no entry signifies $\omega^2 < 0$.

Ra_{stat}	Ra_{osc}	a^2_{osc}	β
690.387	-	-	0.01
723.263	-	-	0.02
756.138	761.819	6.189	0.03
761.398	761.320	6.209	0.0316
789.014	758.936	6.308	0.04
821.889	756.576	6.419	0.05
854.765	754.679	6.523	0.06
887.640	753.197	6.622	0.07
920.516	752.088	6.715	0.08
953.391	751.315	6.803	0.09
986.267	750.848	6.887	0.1

Thus, the combination of two skew-symmetric effects does lead to a non-simple bifurcation surface.

Table 6.5 Critical Rayleigh numbers and oscillatory wave numbers, the stationary wavenumber squared is $\pi^2/2 \approx 4.9348$; $Pr = 25, \gamma = 0.2, Sg = 1, \lambda = 0.001$. Larger values of β.

Ra_{stat}	Ra_{osc}	a^2_{osc}	β
1019.14	750.661	6.968	0.11
1052.02	750.729	7.046	0.12
1084.89	751.033	7.120	0.13
1117.77	751.555	7.193	0.14
1150.64	752.278	7.262	0.15
1183.52	753.189	7.330	0.16
1216.40	754.275	7.396	0.17
1249.27	755.525	7.460	0.18
1282.15	756.929	7.522	0.19
1315.02	758.477	7.583	0.2

Table 6.6 Critical Rayleigh numbers and oscillatory wave numbers, the stationary wavenumber squared is $\pi^2/2 \approx 4.9348$; $Pr = 25, \gamma = 0.2, Sg = 1.4, \lambda = 0.001$.

Ra_{stat}	Ra_{osc}	a^2_{osc}	β
690.387	692.130	7.667	0.01
691.373	691.366	7.666	0.0103
723.263	672.345	7.695	0.02
756.138	659.211	7.747	0.03
789.014	649.515	7.803	0.04
821.889	642.013	7.859	0.05
854.765	636.066	7.914	0.06
887.640	631.293	7.968	0.07
920.516	627.449	8.022	0.08
953.391	624.361	8.076	0.09
986.267	621.905	8.128	0.1

Table 6.7 Critical Rayleigh numbers and oscillatory wave numbers, the stationary wavenumber squared is $\pi^2/2 \approx 4.9348$; $Pr = 25, \gamma = 0.2, Sg = 1.4, \lambda = 0.001$.

Ra_{stat}	Ra_{osc}	a^2_{osc}	β
1019.14	619.987	8.180	0.11
1052.02	618.534	8.232	0.12
1084.89	617.487	8.283	0.13
1117.77	616.800	8.333	0.14
1150.64	616.434	8.383	0.15
1183.52	616.357	8.432	0.16
1216.40	616.541	8.481	0.17
1249.27	616.963	8.529	0.18
1282.15	617.604	8.577	0.19
1315.02	618.445	8.625	0.2

6.4 Kelvin–Voigt fluid of order 2

We now write down the non-dimensional perturbation equations for thermal convection in a Kelvin–Voigt fluid of order two with a Cattaneo equation for the heat flux. In equations (6.13) one takes $L = 2$. The notation is commensurate with that employed in section 6.3 and in Straughan (2022b, section 6), where the analogous problem is analysed with a Fourier heat flux law. In view of the last sentence we replace w_i^1, w_i^2 by W_i, V_i and we replace $\epsilon_1, \epsilon_2, \gamma^1, \gamma^2$ by $\beta, \epsilon, \gamma, \delta$. It should be noted that there are now two extra variables denoted by W_i and V_i accounting for a greater variety of viscoelastic effect in the fluid, whereas there was only one for the Kelvin–Voigt theory of order one. Thus, the non-dimensional perturbation equations for thermal convection in a Kelvin–Voigt fluid of order two with a Cattaneo equation for the heat flux are

$$
\begin{aligned}
&u_{i,t} - \lambda \Delta u_{i,t} + u_j u_{i,j} = -\pi_{,i} + \Delta u_i + R\theta k_i + \beta \Delta W_i + \epsilon \Delta V_i, \\
&u_{i,i} = 0, \\
&Pr(\theta_{,t} + u_i \theta_{,i}) = Rw - q_{i,i}\,, \\
&Sg(q_{i,t} + u_j q_{i,j}) = -q_i - \theta_{,i}\,, \\
&W_{i,t} + \gamma W_i = u_i\,, \\
&V_{i,t} + \delta V_i = u_i\,.
\end{aligned}
\tag{6.22}
$$

The boundary conditions and the steady state are those of section 6.2. Observe that system (6.22) holds on the domain $\mathbb{R}^2 \times \{z \in (0, 1)\}$ for $t > 0$.

We point out that this system contains three sets of skew-symmetric terms. These are the $-q_{i,i}$ and $-\theta_{,i}$ terms, the $\beta \Delta W_i$ and the u_i term in equation $(6.22)_5$, and the $\epsilon \Delta V_i$ and the u_i term in equation $(6.22)_6$. We expect the instability surface in this case to be quite interesting.

Chapter 7
Cattaneo convection with higher gradient Navier–Stokes theory

7.1 CCJ theory, no Mariano effect

A development which is particularly important in the light of the recent interest in microfluid flows, is a generalization of the Navier–Stokes equations to include not only the velocity gradient as a constitutive variable in the stress but also to incorporate the second gradients of the velocity field. This was proposed in a seminal paper by Fried and Gurtin (2006), who developed a complete theory consistent with the principles of modern continuum thermodynamics but they also stressed the importance of finding the correct boundary conditions which is highly non-trivial when second gradients of velocity are involved. This work was further developed by Musesti (2009) who derived the complete constitutive theory. The question of existence of solutions for the Fried–Gurtin–Musesti extended model was investigated by Giusteri et al. (2011) and Degiovanni et al. (2021) who showed that the new model has some highly desirable analytical features. The Fried–Gurtin–Musesti model has been analysed in the context of thermal convection by Straughan (2023d) where the correct boundary conditions of Fried and Gurtin (2006) are studied in detail. We should point out that the question of the correct boundary conditions is a vital one in fluid mechanics. It is imperative to understand the correct boundary conditions to analyse a meaningful physical problem. Straughan (2023b) further analysed the Fried–Gurtin–Musesti theory for convection in a rotating horizontal layer. It was shown that the Fried–Gurtin–Musesti theory leads to very novel and counter intuitive effects. The articles of Straughan (2023d,b) generalized the theory of Fried–Gurtin–Musesti to incorporate thermal effects with a Fourier law for the heat flux in order to be able to examine thermal convection in the higher order Navier–Stokes theory.

The object of the present chapter is to present a model for thermal convection in a horizontal layer when the fluid is a higher gradient Navier–Stokes one of Fried–Gurtin–Musesti type, but to generalize the thermodynamic setting and allow the heat flux to satisfy a Cattaneo equation rather than Fourier's law. We believe this is of interest in its own right, and may be highly relevant in a microfluidic context. We

B. Straughan, *Thermal Convection with a Cattaneo Flux Law*, Advances in Mechanics and Mathematics 53, https://doi.org/10.1007/978-3-031-80885-2_7

commence with a CCJ theory without a Mariano effect in the balance of energy equation.

The general form of equations to represent thermal convection when the heat flux law is of Cattaneo type, generalizing (3.8) are

$$
\begin{aligned}
&\frac{\partial v_i}{\partial t}+v_j\frac{\partial v_i}{\partial x_j}=-\frac{1}{\rho_0}\frac{\partial p}{\partial x_i}+\nu\Delta v_i-\hat{\xi}\Delta^2 v_i+\alpha g k_i T,\\
&\frac{\partial v_i}{\partial x_i}=0,\\
&\frac{\partial T}{\partial t}+v_i\frac{\partial T}{\partial x_i}=-\frac{\partial Q_i}{\partial x_i},\\
&\tau\frac{\mathcal{D}Q_i}{\mathcal{D}t}+Q_i=-k\frac{\partial T}{\partial x_i},
\end{aligned}
\tag{7.1}
$$

where the terms are as in (2.27) although the new term in $\Delta^2 v_i$ has a coefficient $\hat{\xi}$ which is usually referred to as a hyperviscosity coefficient. The derivate $\mathcal{D}Q_i/\mathcal{D}t$ can be the material derivative or one of the objective derivatives studied in chapter 3. In this section we assume this derivative is the material derivative and so the relevant equations for Cattaneo–Christov–Jordan theory are

$$
\begin{aligned}
&\frac{\partial v_i}{\partial t}+v_j\frac{\partial v_i}{\partial x_j}=-\frac{1}{\rho_0}\frac{\partial p}{\partial x_i}+\nu\Delta v_i-\hat{\xi}\Delta^2 v_i+\alpha g k_i T,\\
&\frac{\partial v_i}{\partial x_i}=0,\\
&\frac{\partial T}{\partial t}+v_i\frac{\partial T}{\partial x_i}=-\frac{\partial Q_i}{\partial x_i},\\
&\tau\Big(\frac{\partial Q_i}{\partial t}+v_j\frac{\partial Q_i}{\partial x_j}\Big)+Q_i=-k\frac{\partial T}{\partial x_i}.
\end{aligned}
\tag{7.2}
$$

The boundary conditions we consider at the beginning need to be more than those in chapter 3 because the order of the momentum equation (7.2) is two derivatives higher. Various possible boundary conditions are discussed by Fried and Gurtin (2006) and adapted to thermal convection by Straughan (2023d). We here only consider those for strong adherence and these are

$$
\begin{aligned}
&v_i=0,\qquad \frac{\partial v_i}{\partial n}=0,\qquad z=0,d;\\
&T=T_L,\quad z=0,\qquad T=T_U,\quad z=d,
\end{aligned}
\tag{7.3}
$$

where the derivative $\partial/\partial n$ refers to the derivative in the direction outward of the unit normal to the fluid layer, T_L, T_U are constants with $T_L > T_U$.

The steady solution to the model (7.2) corresponding to the boundary conditions (7.3) is

$$
\bar{v}_i\equiv 0,\qquad \bar{T}=-\beta z+T_L,\qquad \bar{\mathbf{Q}}=(0,0,k\beta). \tag{7.4}
$$

In order to analyse stability the perturbation quantities are introduced as in chapter 3. The non-dimensional variables are as in chapter 3 with the addition of the parameter ξ which is a non-dimensional version of the hyperviscosity coefficient. The non-dimensional perturbation boundary conditions are

$$u_i = 0, \quad \frac{\partial u_i}{\partial n} = 0, \quad \theta = 0, \qquad z = 0, 1, \tag{7.5}$$

with the solution u_i, θ, q_i, π satisfying a periodic pattern in the x, y directions.

The fully nonlinear, non-dimensional perturbation equations may be written for the CCJ system as,

$$\begin{aligned} &u_{i,t} + u_j u_{i,j} = -\pi_{,i} + \Delta u_i - \xi \Delta^2 u_i + R k_i \theta, \\ &u_{i,i} = 0, \\ &Pr(\theta_{,t} + u_i \theta_{,i}) = Rw - q_{i,i}, \\ &Sg(q_{i,t} + u_j q_{i,j}) = -\theta_{,i} - q_i, \end{aligned} \tag{7.6}$$

where ξ is a non-dimensional version of the hyperviscosity parameter.

To determine the linear instability boundary we linearize (7.6) and use a representation $u_i = u_i(\mathbf{x})e^{\sigma t}$, $\theta = \theta(\mathbf{x})e^{\sigma t}$, $q_i = q_i(\mathbf{x})e^{\sigma t}$, $\pi = \pi(\mathbf{x})e^{\sigma t}$. The pressure is removed by taking curl twice of the linearized version of $(7.6)_1$. Thus, one may arrive at the system of linear instability equations

$$\begin{aligned} &-\sigma \Delta w = -\Delta^2 w + \xi \Delta^3 w - R \Delta^* \theta, \\ &Pr\sigma\theta = Rw - h, \\ &Sg\sigma h = -\Delta\theta - h, \end{aligned} \tag{7.7}$$

where $h = q_{i,i}$. One then uses a normal mode analysis and with a being the wavenumber and $D = d/dz$, one has to solve the eigenvalue problem

$$\begin{aligned} &-\sigma(D^2 - a^2)W = -(D^2 - a^2)^2 W + \xi(D^2 - a^2)^3 W + Ra^2\Theta, \\ &Pr\sigma\Theta = Rw - h, \\ &Sg\sigma H = -(D^2 - a^2)\Theta - H, \end{aligned} \tag{7.8}$$

for $z \in (0, 1)$, where a is the wavenumber and where $W(z), \Theta(z)$ and $H(z)$ are the $z-$dependent parts of w, θ and h. System (7.8) has to be solved for the eigenvalue(s) σ subject to the boundary conditions

$$W = DW = D^2 W = \Theta = 0, \qquad z = 0, 1, \tag{7.9}$$

cf. Straughan (2023d).

The eigenvalue problem (7.8) and (7.9) needs to be solved numerically. Typically ξ, Pr and Sg are very small and this results in a difficult eigenvalue problem to resolve numerically since the smallness of the coefficients make it almost singular. Even a robust method like the Chebyshev–tau one, cf. Dongarra et al. (1996), presents a challenge to find accurate numerical values, especially when σ is complex, cf. Straughan (2023b).

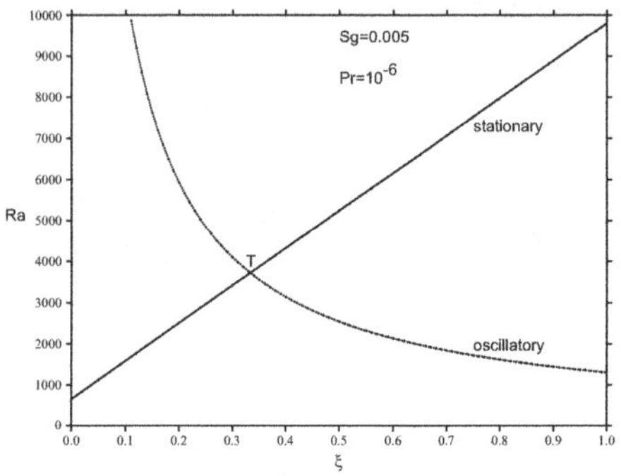

Fig. 7.1 Detail of transition behaviour between stationary convection and oscillatory convection. The transition occurs at Ra = 3723, ξ = 0.333.

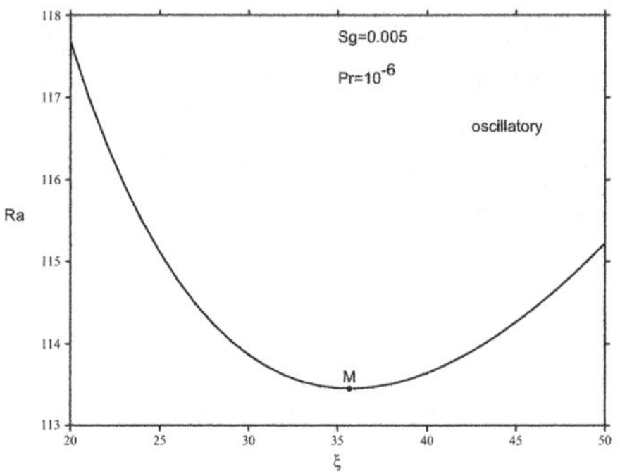

Fig. 7.2 Detail of oscillatory convection curve, showing the minimum value at Ra = 113.457, ξ = 35.67.

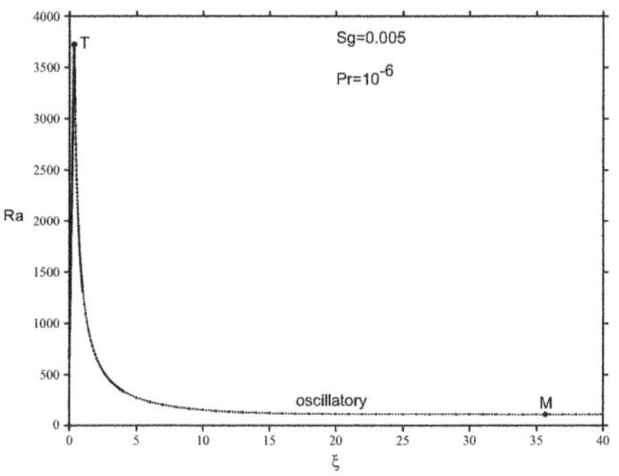

Fig. 7.3 Ra instability threshold. Stationary convection from ξ = 0 to ξ = 0.333 and oscillatory convection thereafter. The transition from stationary convection to oscillatory convection is at Ra = 3723, ξ = 0.333, while the minimum value on the oscillatory convection curve is at Ra = 113.457, ξ = 35.67.

In this chapter we proceed with a set of illustrative boundary conditions, cf. (Straughan, 2023d, section 6.5), which allow one to use an analytical method further. One replaces boundary conditions (7.9) by the illustrative boundary conditions, (Straughan, 2023d, section 6.5),

$$W = D^2W = D^4W = \Theta = 0, \qquad z = 0, 1. \tag{7.10}$$

One may then solve (7.8), (7.10) as sin series to find the linear instability thresholds are given by, for stationary convection, ($\sigma \in \mathbb{R}$),

$$R^2 = \frac{\Lambda^3}{a^2} + \xi \frac{\Lambda^4}{a^2}, \tag{7.11}$$

and for oscillatory convection, ($\sigma \in \mathbb{C}$),

$$R^2 = \frac{Pr}{Sg}\frac{\Lambda}{a^2} + \frac{1}{Sg^2(1+\xi\Lambda)}\frac{\Lambda}{a^2} + \frac{Pr}{Sg}(1+\xi\Lambda)\frac{\Lambda^2}{a^2}, \tag{7.12}$$

where $\Lambda = \pi^2 + a^2$.

The critical Rayleigh number instability thresholds are now found by minimizing (7.11) and (7.12) in a^2 numerically. Details of the stationary convection curves may be found in Straughan (2023d) for various values of ξ. One may observe from (7.12) that when Sg is large enough oscillatory convection will be the dominant mechanism of instability.

We present numerical results of the critical stationary and oscillatory convection curves with a value of the Prandtl number, $Pr = 10^{-6}$. In convection in stellar atmospheres the Prandtl number is typically in the region $10^{-9} \leq Pr \leq 10^{-5}$, see Käpylä (2021), Garaud (2021).

The instability curves depend strongly on the value of Sg. When $Sg = 0.01$ we find Ra_{osc} is always less than Ra_{stat} and then the minimum value as one varies ξ is

$$R^2_{osc} = 40.0765, \quad a^2_{osc} = 19.9, \qquad \text{when} \qquad \xi = 25.$$

For values of $Sg = 10^{-3}, 10^{-4}, 10^{-5}$ we find stationary convection occurs for ξ small enough then when ξ is large enough oscillatory convection commences and this reaches a minimum value for a certain value of ξ and then the oscillatory convection Rayleigh number continues to increase with further increase in ξ. For example, when $Sg = 10^{-3}$, the minimum in ξ is

$$R^2_{osc} = 40.0765, \quad a^2_{osc} = 19.9, \qquad \text{when} \qquad \xi = 25,$$

when $Sg = 10^{-4}$, the minimum in ξ is

$$R^2_{osc} = 40882.7, \quad a^2_{osc} = 20.2, \qquad \text{when} \qquad \xi = 247,$$

and when $Sg = 10^{-5}$, the minimum in ξ is

$$R^2_{osc} = 1.35346 \times 10^6, \quad a^2_{osc} = 21.2, \qquad \text{when} \qquad \xi = 749.$$

Details of the instability thresholds are provided in figures 7.1, 7.2 and 7.3, for the value of $Sg = 0.005$. In figure 7.3 we see that stationary convection occurs for ξ small, this threshold rises rapidly from $Ra_{stat} \approx 657.5$ at $\xi = 0$ to $Ra_{stat} \approx 3723$ where

$\xi = 0.333$. There is then a transition, point T in figure 7.3, and oscillatory convection occurs. The oscillatory convection curve decreases rapidly as ξ increases and reaches a minimum, M, at $Ra_{osc} = 113.457$ when $\xi = 35.67$. As ξ is increased further the oscillatory convection Rayleigh number increases. Figure 7.1 shows details of the transition from stationary to oscillatory convection with $\xi \in [0, 1]$. Details of the minimum on the oscillatory convection branch are shown in figure 7.2. For the values of $Sg = 10^{-5}, 10^{-4}, 10^{-3}$ the behaviour of the Rayleigh number threshold is qualitatively similar to that shown in figures 7.1 - 7.3, although the transition point and the minimum on the oscillatory convection curve are very different.

The behaviour of the Rayleigh number threshold in figure 7.3 is very interesting in the region between the points on the graph marked T and M. In this region the Rayleigh number threshold is decreasing as ξ increases. However, the operator $-\xi\Delta^2$ in the governing equations is a strongly positive operator which would normally lead to strong stabilization as ξ is increased. In this case we observe strong destabilization by oscillatory convection. This would appear to indicate that there is a very interesting non–intuitive behaviour taking place with an interaction of the Cattaneo theory and the hyperviscosity term. A similar behaviour was witnessed by Straughan (2023b) with the hyperviscosity when the layer was rotating, except there the curve which showed a decrease was the stationary convection one.

As was seen in chapter 3 in the Navier–Stokes case, one may show that a standard energy stability analysis does not follow from equations (7.6) due to the lack of dissipation in the temperature. We now examine the model of this section but include a Mariano effect in the heat flux.

7.2 CCJ theory with Mariano effect

We now analyse thermal convection in a horizontal layer of Fried–Gurtin–Musesti fluid with the CCJ theory for the heat flux, and we incorporate the extra flux of Mariano (2017).

The basic equations for this problem have form

$$
\begin{aligned}
&\frac{\partial v_i}{\partial t} + v_j \frac{\partial v_i}{\partial x_j} = -\frac{1}{\rho_0}\frac{\partial p}{\partial x_i} + \nu\Delta v_i - \xi\Delta^2 v_i + \alpha g k_i T, \\
&\frac{\partial v_i}{\partial x_i} = 0, \\
&\frac{\partial T}{\partial t} + v_i \frac{\partial T}{\partial x_i} = -\frac{\partial Q_i}{\partial x_i} + \hat{\zeta}\Delta T, \\
&\tau\Big(\frac{\partial Q_i}{\partial t} + v_j \frac{\partial Q_i}{\partial x_j}\Big) + Q_i = -k\frac{\partial T}{\partial x_i},
\end{aligned}
\qquad (7.13)
$$

and the boundary conditions now are

$$v_i = 0, \quad \frac{\partial v_i}{\partial n} = 0, \qquad z = 0, d;$$
$$T = T_L, \quad z = 0, \qquad T = T_U, \quad z = d. \tag{7.14}$$

The steady solution whose stability is under investigation is given by (7.4).

With ζ denoting a non-dimensional form of $\hat{\zeta}$, the non-dimensional perturbation equations arising from (7.13) are

$$\begin{aligned} &u_{i,t} + u_j u_{i,j} = -\pi_{,i} + \Delta u_i - \xi \Delta^2 u_i + R k_i \theta, \\ &u_{i,i} = 0, \\ &Pr(\theta_{,t} + u_i \theta_{,i}) = Rw - q_{i,i} + \zeta \Delta \theta, \\ &Sg(q_{i,t} + u_j q_{i,j}) = -\theta_{,i} - q_i, \end{aligned} \tag{7.15}$$

and the boundary conditions are

$$u_i = 0, \quad \frac{\partial u_i}{\partial n} = 0, \quad \theta = 0, \qquad z = 0, 1, \tag{7.16}$$

with the solution u_i, θ, q_i, π satisfying a periodic pattern in the x, y directions.

To solve the linear instability problem for (7.15), (7.16), one linearizes and removes the pressure term. Then, one employs a temporal description like $e^{\sigma t}$ to see that one must now solve the eigenvalue problem

$$\begin{aligned} &-\Delta w = -\Delta^2 w - R\Delta^* \theta + \xi \Delta^3 w, \\ &\sigma Pr \theta = Rw - q_{i,i} + \zeta \Delta \theta, \\ &\sigma Sg q_i + q_i = -\theta_{,i}, \end{aligned} \tag{7.17}$$

subject to the boundary conditions

$$w = 0, \quad \frac{\partial w}{\partial z} = 0, \quad \frac{\partial^2 w}{\partial z^2} = 0, \quad \theta = 0, \qquad z = 0, 1. \tag{7.18}$$

In general, it would appear necessary to solve this eigenvalue problem by numerical means. A suitable method is likely to be the D^2 Chebyshev tau method, cf. Dongarra et al. (1996).

One may develop an energy stability analysis from the fully nonlinear equations (7.15). This would entail multiplying $(7.15)_1$ by u_i, $(7.15)_3$ by θ, $(7.15)_4$ by q_i, and integrating each in turn over a period cell V and employing boundary conditions (7.16). In this way one could arrive at an energy equation of form

$$\frac{dE}{dt} = RI - D, \tag{7.19}$$

where the energy function E is

$$E = \frac{1}{2}\|\mathbf{u}\|^2 + \frac{Pr}{2}\|\theta\|^2 + \frac{Sg}{2}\|\mathbf{q}\|^2,$$

the production term I is

$$I = 2(\theta, w),$$

while the dissipation is

$$D = \|\nabla \mathbf{u}\|^2 + \xi\|\Delta \mathbf{u}\|^2 + \zeta\|\nabla\theta\|^2 + \|\mathbf{q}\|^2.$$

One may then define

$$\frac{1}{R_E} = \max_H \frac{I}{D},$$

where H is the space of admissible solutions. The Euler–Lagrange equations for this maximum are

$$\begin{aligned} \Delta u_i - \xi\Delta^2 u_i + R_E\theta k_i &= \lambda_{,i}, \\ u_{i,i} &= 0, \\ \zeta\Delta\theta + R_E w &= 0, \end{aligned}$$

where λ is a Lagrange multiplier. This eigenvalue problem must be solved for R_E subject to the boundary conditions (7.16).

We now present details of a solution to the linear instability and nonlinear energy stability problems where idealized boundary conditions are employed.

For such idealized boundary conditions one may show

$$R_{stat}^2 = (\zeta + 1)\left(\frac{\Lambda^3}{a^2} + \xi\frac{\Lambda^4}{a^2}\right), \tag{7.20}$$

and

$$\begin{aligned} R_{osc}^2 = &\frac{\zeta\Lambda^2 + XPr}{Sga^2} + \frac{Pr\Lambda(\zeta\Lambda^2 + XPr)}{Sg^2a^2(\zeta\Lambda^2 + PrX)} \\ &+ \frac{\zeta\Lambda X}{a^2} + \frac{\zeta Sg\Lambda^2 + Pr\Lambda}{Sg^2a^2(\zeta + Pr[1 + \xi\Lambda])}, \end{aligned} \tag{7.21}$$

where $X = \Lambda^2 + \xi\Lambda^3$. To find the critical Rayleigh number for stationary convection one must minimize (7.20) in a^2, whereas to determine the critical Rayleigh number for oscillatory convection it is necessary to minimize (7.21) in a^2. The second and fourth terms on the right of (7.21) are likely to lead to interesting behaviour.

With ideal boundary conditions the energy stability boundary is given by

$$R_{stat}^2 = \zeta\left(\frac{\Lambda^3}{a^2} + \xi\frac{\Lambda^4}{a^2}\right). \tag{7.22}$$

Again, the critical value of the energy stability Rayleigh number is found by minimizing (7.22) in a^2.

We observe that

$$R_{stat}^2 \geq R_E^2 \qquad \text{and} \qquad R_{osc}^2 \geq R_E^2,$$

as they should be.

7.3 Cattaneo–Morro theory with Mariano effect

In this section we return to analyse Cattaneo–Morro–Mariano theory as in section 4.4, but now in the context of the Fried–Gurtin–Musesti equations rather than Navier–Stokes. Again, we investigate thermal convection in a layer of incompressible fluid heated from below. The basic equations may be written as (7.1) but with a Mariano extra flux and incorporating a Green–Naghdi dissipation in the heat flux. The steady state solution is again as in (7.4). If we adopt Cattaneo–Morro–Mariano theory then the appropriate perturbation equations which replace (7.15) have the form

$$\begin{aligned} &u_{i,t} + u_j u_{i,j} = -\pi_{,i} + \Delta u_i - \xi \Delta^2 u_i + R k_i \theta, \\ &u_{i,i} = 0, \\ &Pr(\theta_{,t} + u_i \theta_{,i}) = Rw - q_{i,i} + \zeta \Delta \theta\,, \\ &Sg(q_{i,t} + u_j q_{i,j} - \omega_{ij} q_j + \gamma d_{ij} q_j) = -\theta_{,i} - q_i + \epsilon \Delta q_i + \frac{SgR}{Pr}(\omega_{i3} - \gamma d_{i3}), \end{aligned} \tag{7.23}$$

and the boundary conditions we consider are

$$u_i = 0, \qquad \frac{\partial u_i}{\partial n} = 0\,, \qquad \theta = 0, \qquad q_i = 0, \qquad z = 0, 1, \tag{7.24}$$

with the solution u_i, θ, q_i, π satisfying a periodic pattern in the x, y directions.

In this chapter we only discuss energy stability for (7.23) and (7.24). Thus, we multiply $(7.23)_1$ by u_i, multiply $(7.23)_3$ by θ, multiply $(7.23)_4$ by q_i, and we integrate each resulting equation over a period cell V and use the boundary conditions to derive the energy identities

$$\frac{d}{dt}\frac{1}{2}\|\mathbf{u}\|^2 = -\|\nabla \mathbf{u}\|^2 - \xi \|\Delta \mathbf{u}\|^2 + R(\theta, w), \tag{7.25}$$

and

$$\frac{d}{dt}\frac{Pr}{2}\|\theta\|^2 = -\zeta \|\nabla \theta\|^2 - (q_{i,i}, \theta) + R(\theta, w), \tag{7.26}$$

together with

$$\begin{aligned} \frac{d}{dt}\frac{Sg}{2}\|\mathbf{q}\|^2 = &- \|\mathbf{q}\|^2 - \epsilon \|\nabla \mathbf{q}\|^2 - (\theta_{,i}, q_i) \\ &- \gamma \int_V d_{ij} q_i q_j \, dx + \frac{SgR}{Pr}(\omega_{i3}, q_i) - \frac{\gamma SgR}{Pr}(d_{i3}, q_i). \end{aligned} \tag{7.27}$$

We add the last three equations after multiplying $(7.25)_1$ by a positive coupling parameter λ to arrive at an energy equation

$$\begin{aligned}\frac{dE}{dt} =& R(1+\lambda)(\theta,w)+\frac{SgR}{Pr}(\omega_{i3},q_i)-\frac{\gamma SgR}{Pr}(d_{i3},q_i)\\ &-\|\mathbf{q}\|^2-\epsilon\|\nabla\mathbf{q}\|^2-\lambda\|\nabla\mathbf{u}\|^2-\lambda\xi\|\Delta\mathbf{u}\|^2\\ &-\gamma\int_V d_{ij}q_iq_j\,dx.\end{aligned}\tag{7.28}$$

There are various ways one may proceed at this point. We here use the arithmetic-geometric mean inequality to arrive at the inequality

$$\frac{dE}{dt}\le R(1+\lambda)(\theta,w)-D-\gamma\int_V d_{ij}q_iq_j\,dx,\tag{7.29}$$

where the energy function is

$$E(t)=\frac{\lambda}{2}\|\mathbf{u}\|^2+\frac{Pr}{2}\|\theta\|^2+\frac{Sg}{2}\|\mathbf{q}\|^2,\tag{7.30}$$

and the dissipation is given by

$$D(t)=\zeta\|\nabla\theta\|^2+A\|\nabla\mathbf{u}\|^2+\lambda\xi\|\Delta\mathbf{u}\|^2+B\|\mathbf{q}\|^2+\epsilon\|\nabla\mathbf{q}\|^2\,.\tag{7.31}$$

The constants A and B have form

$$A=\lambda-(1+\gamma^2)\,\frac{SgR}{Pr},\qquad B=1-\frac{SgR}{2Pr},$$

where we are restricting attention to Sg relatively small so that $A>0$, $B>0$.

One may develop an energy stability theory from (7.29) in a normal way using the production term $R(1+\lambda)(\theta,w)$ and the dissipation function D. The idea is to select λ in such a way as to maximize the energy stability Rayleigh number. However, as in section 4.4, we need to handle the cubic nonlinear term involving $d_{ij}q_iq_j$. To do this we write

$$\begin{aligned}-\gamma\int_V d_{ij}q_iq_j\,dx =& -\gamma\int_V u_{i,j}q_iq_j\,dx\\ =& \gamma\int_V u_iq_iq_{j,j}\,dx+\gamma\int_V u_iq_jq_{i,j}\,dx\,,\end{aligned}\tag{7.32}$$

where we have integrated by parts and used the boundary conditions (7.24). We next observe that

$$\int_V (q_{i,i})^2dx\le 3\|\nabla\mathbf{q}\|^2.\tag{7.33}$$

From (7.32) we derive

$$\begin{aligned}-\gamma\int_V d_{ij}q_iq_j\,dx \le& \gamma\sup_V|\mathbf{u}|\,\|\mathbf{q}\|\,\|q_{i,i}\|+\gamma\sup_V|\mathbf{u}|\,\|\mathbf{q}\|\,\|\nabla\mathbf{q}\|\,,\\ \le& \gamma c(\|\mathbf{q}\|\,\|\Delta\mathbf{u}\|\,\|q_{i,i}\|+\|\mathbf{q}\|\,\|\Delta\mathbf{u}\|\,\|\nabla\mathbf{q}\|),\end{aligned}\tag{7.34}$$

where we have employed the Sobolev inequality

$$\sup_V |\mathbf{u}| \leq c||\Delta\mathbf{u}||.$$

A value for the constant c may be found in (Galdi and Straughan, 1985b, equation (A15)). Thus, using (7.33) in (7.34) one obtains

$$-\gamma \int_V d_{ij}q_iq_j\,dx \leq \frac{5\gamma c}{2}\,||\Delta\mathbf{u}||\,||\mathbf{q}||\,||\nabla\mathbf{q}||. \tag{7.35}$$

Inequality (7.35) allows one to develop an energy stability analysis for a three-dimensional cell, V, since from the forms for E and D in (7.30) and (7.31), there is a constant k such that

$$-\gamma \int_V d_{ij}q_iq_j\,dx \leq kDE^{1/2}. \tag{7.36}$$

Then, one may use (7.36) in (7.29) to find

$$\frac{dE}{dt} \leq R(1+\lambda)(\theta, w) - D + kDE^{1/2}. \tag{7.37}$$

A conditional nonlinear energy stability analysis may be developed from inequality (7.37) even in the three-dimensional case. This is a situation where the hyperviscosity term in the Fried–Gurtin–Musesti equations allows one to control the nonlinear term arising from use of the objective derivative in the Cattaneo–Morro–Mariano theory. This is unlike the situation in section 4.4 where it was necessary to employ the Navier–Stokes–Voigt equations to be able to control the nonlinearity.

Chapter 8
Cattaneo convection with double diffusion

8.1 Double diffusion in a Darcy porous material

In this chapter we address the problem of thermal convection in a layer of porous material saturated with an incompressible fluid. The heat flux satisfies a Cattaneo equation, but the fluid also contains a concentration of a dissolved salt. We commence with a layer of Darcy porous material which is saturated with an incompressible fluid containing a dissolved salt, as analysed by Straughan (2011b).

The governing equations consist of Darcy's law which when coupled with a Boussinesq approximation for the density in the body force take the form

$$\frac{\partial p}{\partial x_i} = -\frac{\mu}{K} v_i + \rho_0 \alpha g T k_i - \rho_0 \alpha_s g C k_i, \tag{8.1}$$

where $p, \mu, K, \rho_0, \alpha, g, \alpha_s$ are the pressure, the dynamic viscosity of the fluid, the permeability, the constant fluid density, the coefficient of thermal expansion of the fluid, gravity, and the solute expansion coefficient. The functions v_i, T and C are the pore averaged velocity, the temperature, and the concentration of salt in the porous medium, and $\mathbf{k} = (0, 0, 1)$. The fluid is incompressible and so

$$\frac{\partial v_i}{\partial x_i} = 0. \tag{8.2}$$

The equation for the concentration of the salt is

$$\phi \frac{\partial C}{\partial t} + v_i \frac{\partial C}{\partial x_i} = \phi k_C \Delta C, \tag{8.3}$$

where ϕ is porosity, k_C is the salt diffusion coefficient, and Δ is the Laplacian. By writing separate equations for the temperature in the porous solid and for the fluid in the pores Straughan (2011b) deduces that the governing equations for the temperature and heat flux, Q_i, have form

B. Straughan, *Thermal Convection with a Cattaneo Flux Law*, Advances in Mechanics and Mathematics 53, https://doi.org/10.1007/978-3-031-80885-2_8

$$\frac{1}{M}\frac{\partial T}{\partial t}+v_i\frac{\partial T}{\partial x_i}=-\frac{\partial Q_i}{\partial x_i}\,,\tag{8.4}$$

and

$$\tau\frac{\partial Q_i}{\partial t}+\tau_f\left(v_j\frac{\partial Q_i}{\partial x_j}-Q_j\frac{\partial v_i}{\partial x_j}\right)=-Q_i-\kappa\frac{\partial T}{\partial x_i}\,,\tag{8.5}$$

where M represents a ratio of density and specific heats, τ is a relaxation time for the porous medium and τ_f is an analogous quantity for the fluid. In (8.5) the Cattaneo–Christov derivative has been employed.

Equations, (8.1), (8.2), (8.3), (8.4) and (8.5) hold on $\mathbb{R}^2\times\{z\in(0,d)\}$ for $t>0$. On the boundaries $z=0,d$ the boundary conditions are

$$\begin{aligned}&T=T_L,\quad C=C_L,\quad z=0;\qquad T=T_U,\quad C=C_U,\quad z=d;\\&v_3=w=0,\qquad z=0,d.\end{aligned}\tag{8.6}$$

Here $T_L>T_U$, $C_L>C_U$, they are constants and in this situation the layer is heated from below and simultaneously salted from below. The steady solution is

$$\begin{aligned}&\bar{v}_i\equiv0,\quad \bar{C}=-\beta_s z+C_L\,,\quad \beta_s=\frac{C_L-C_U}{d}\,,\\&\bar{T}=-\beta z+T_L\,,\quad \beta=\frac{T_L-T_U}{d}\quad \bar{Q}_3=\kappa\beta.\end{aligned}\tag{8.7}$$

The steady pressure $\bar{p}(z)$ is found from (8.1). Perturbations u_i,π,θ,q_i and φ are introduced to the steady solution and these are non-dimensionalized as in Straughan (2011b). This leads to the following system of non-dimensional perturbation equations,

$$\begin{aligned}&\frac{\partial\pi}{\partial x_i}=-u_i+R\theta k_i\underline{-R_s\varphi k_i},\\&\frac{\partial u_i}{\partial x_i}=0,\\&\varepsilon\frac{\partial\varphi}{\partial t}+Le u_i\frac{\partial\varphi}{\partial x_i}=\underline{R_s w}+\Delta\varphi,\\&\frac{\partial\theta}{\partial t}+u_i\frac{\partial\theta}{\partial x_i}=Rw\underline{\underline{-\frac{\partial q_i}{\partial x_i}}},\\&C\frac{\partial q_i}{\partial t}+\hat{\tau}_f\left(u_j\frac{\partial q_i}{\partial x_j}-q_j\frac{\partial u_i}{\partial x_j}\right)=2C_fR\frac{\partial u_i}{\partial z}-q_i\underline{\underline{-\frac{\partial\theta}{\partial x_i}}}.\end{aligned}\tag{8.8}$$

Equations (8.8) hold on the domain $\mathbb{R}^2\times\{z\in(0,1)\}$ for $t>0$, and the solution satisfies the boundary conditions

$$w=u_in_i=u_3=0,\quad \theta=\varphi=0,\qquad z=0,1,\tag{8.9}$$

and additionally the solution satisfies a plane tiling periodic pattern in (x,y). In (8.8) Le is the Lewis number, $\varepsilon=\phi M Le$, $C=M\kappa\tau/d^2$ is a Cattaneo number, $\hat{\tau}_f$ is a non-dimensional version of τ_f, C_f is a fluid Cattaneo number, $R^2 = Ra$ is the Rayleigh number, and R_s^2 is the salt Rayleigh number.

Even in the case of Fourier's law the equations for thermosolutal convection in a porous medium corresponding to (8.8) contain a skew-symmetric contribution to the linear operator. This is also true here and is manifest by the single underlined terms in $(8.8)_1$ and $(8.8)_3$, namely the $-R_s\varphi k_i$ and $R_s w$ terms. This skew-symmetric contribution leads to oscillatory convection being the dominant mode of convective motion for certain parameter regimes. Here, however, due to the Cattaneo equation there is a second skew-symmetric operator part to the linear operator. This consists of the double underlined terms in $(8.8)_4$ and $(8.8)_5$, namely the terms $-\partial q_i/\partial x_i$ and $-\partial\theta/\partial x_i$. Again, this second skew-symmetric contribution leads to oscillatory convection and we expect the bifurcation to convection picture to be richer than in the Fourier case.

To analyse instability equations (8.8) are linearized, we set $\Gamma = q_{i,i}$, and we write $w = e^{\sigma t}w(\mathbf{x})$ with analogous forms for θ, π, φ and Γ. Then the linearized system may be reduced to the equations

$$\begin{aligned}
&\Delta w - R\Delta^*\theta + R_s\Delta^*\varphi = 0,\\
&-\Gamma + Rw = \sigma\theta,\\
&\Delta\theta + \Gamma = -\sigma C\Gamma,\\
&\Delta\varphi + R_s w = \sigma\varepsilon\varphi,
\end{aligned} \tag{8.10}$$

with the boundary conditions

$$w = 0, \quad \theta = 0, \quad \varphi = 0, \qquad z = 0, 1. \tag{8.11}$$

The solution is assumed periodic in x, y and the planform f satisfies $\Delta^* f = -a^2 f$, with a being the wavenumber, cf. Chandrasekhar (1981, pages 43–52). By writing $w = W(z)f(x, y)$ with a similar representation for θ, φ and Γ, these equations reduce to

$$\begin{aligned}
&(D^2 - a^2)W + Ra^2\Theta - R_s^2 a^2\Phi = 0,\\
&(D^2 - a^2)\Theta + \Gamma = -\sigma C\Gamma,\\
&(D^2 - a^2)\Phi + R_s W = \sigma\varepsilon\Phi,\\
&-\Gamma + RW = \sigma\Theta,
\end{aligned} \tag{8.12}$$

where $D = d/dz$, and they have to be solved for the eigenvalues σ_n, subject to the boundary conditions

$$W = 0, \quad \Theta = 0, \quad \Phi = 0, \qquad z = 0, 1. \tag{8.13}$$

Straughan (2011b) finds the stationary convection boundary is given by

$$R^2 = R_s^2 + \frac{\Lambda^2}{a^2}, \tag{8.14}$$

where $\Lambda = \pi^2 + a^2$. The function R^2 is minimized in a to yield

$$R^2 = R_s^2 + 4\pi^2 \qquad \text{and} \qquad a_c = \pi,$$

a_c being the value of a which yields a minimum to R^2 in (8.14).

Straughan (2011b) also finds the oscillatory convection boundary numerically. The instability threshold is a two-dimensional surface in the three-dimensional space of R^2, R_s^2, C. For $R_s = 0$ the instability threshold is the straight line $R^2 = 4\pi^2$ for $0 \leq C \leq 1/4\pi^2$ and then takes the form $R^2 = 1/C$ for $C > 1/4\pi^2$. In the oscillatory region the critical wavenumber a has infinite value, cf. Straughan (2010a). For $C = 0$ the instability curve starts at $R^2 = 4\pi^2$ and increases linearly in R_s^2 as R_s^2 increases. It increases until $R_s^2 = 4\pi^2/(\varepsilon - 1)$ and thereafter is another straight line with slope $1/\varepsilon$. In the region $R_s^2 > 0, C > 0$ the surface is complicated. For example, for $\varepsilon = 10$, there is an interval $R_s^2 \in (R_1^2, R_2^2)$ where stationary convection is witnessed for small C, then oscillatory convection is observed until a second transition in C is reached. The surface is described in more detail in Straughan (2011b).

8.2 Vadasz inertia effect

In Straughan (2011b) the fluid inertia was neglected in the equation for Darcy's law, (8.1). In this section we include this effect and find it has a strong influence on the instability behaviour. The only change to equations (8.1) - (8.5) is to modify (8.1) as follows

$$\mathcal{P}\frac{\partial v_i}{\partial t} = -\frac{\partial p}{\partial x_i} - \frac{\mu}{K}\, v_i + \rho_0 \alpha g T k_i - \rho_0 \alpha_s g C k_i, \tag{8.15}$$

where $\mathcal{P}$ is the inertia coefficient. The boundary conditions are the same as in section 8.1, and the steady solution is again (8.7). Perturbations are introduced as in section 8.1 and now the non-dimensional perturbation equations take the form

$$\begin{aligned}
&\frac{1}{Va}\frac{\partial u_i}{\partial t} = -\frac{\partial \pi}{\partial x_i} - u_i + R\theta k_i \underline{- R_s \varphi k_i},\\
&\frac{\partial u_i}{\partial x_i} = 0,\\
&\varepsilon\frac{\partial \varphi}{\partial t} + Le u_i \frac{\partial \varphi}{\partial x_i} = \underline{R_s w} + \Delta\varphi,\\
&\frac{\partial \theta}{\partial t} + u_i \frac{\partial \theta}{\partial x_i} = Rw \underline{\underline{- \frac{\partial q_i}{\partial x_i}}},\\
&C\frac{\partial q_i}{\partial t} + \hat{\tau}_f\Big(u_j \frac{\partial q_i}{\partial x_j} - q_j \frac{\partial u_i}{\partial x_j}\Big) = 2C_f R \frac{\partial u_i}{\partial z} - q_i \underline{\underline{- \frac{\partial \theta}{\partial x_i}}}.
\end{aligned} \tag{8.16}$$

where Va is the Vadasz number, cf. Vadasz (1998), Straughan (2000). The boundary conditions are again (8.9). There are again two skew-symmetric parts to the linear operator in equations (8.16), but now there is additionally another growth rate term σ which arises due to $\partial u_i/\partial t$. This contributes to a yet richer structure of the instability bifurcation surface for the initiation of convective motion.

Equations (8.16) may be linearized and further reduced to the form

$$\begin{aligned}
-\frac{\sigma}{Va}\Delta w &= \Delta w - R\Delta^*\theta + R_s\Delta^*\varphi,\\
\varepsilon\sigma\varphi &= R_s w + \Delta\varphi,\\
\sigma\theta &= Rw - \Gamma,\\
C\sigma\Gamma &= -\Gamma - \Delta\theta.
\end{aligned} \tag{8.17}$$

The stationary convection boundary is the same as in section 8.1, so

$$R_{stat}^2 = R_s^2 + \frac{\Lambda^2}{a^2}\,.$$

One may employ a normal mode analysis to (8.17) to find that the oscillatory convection boundary satisfies the equation

$$a_1\sigma^4 + a_2\sigma^3 + a_3\sigma^2 + a_4\sigma + \Lambda R_s^2 a^2 + \Lambda^3 - R^2a^2\Lambda = 0, \tag{8.18}$$

where $a_1,\ldots,a_4$ are given by

$$\begin{aligned}
a_1 &= \frac{\Lambda C\varepsilon}{Va}, \qquad a_2 = \Lambda C\Big(\varepsilon + \frac{\Lambda}{Va}\Big) + \frac{\Lambda\varepsilon}{Va},\\
a_3 &= \frac{\Lambda^2\varepsilon}{Va} + \Lambda^2 C + \epsilon\Lambda + \frac{\Lambda^2}{Va} + R_s^2a^2C - R^2a^2\varepsilon C,\\
a_4 &= \Lambda^2\Big(\varepsilon + \frac{\Lambda}{Va}\Big) + \Lambda^2 + R_s^2a^2 - R^2a^2(\varepsilon + \Lambda C).
\end{aligned}$$

Take the real and imaginary parts of (8.18) and then one may show

$$R^2 = \frac{Y - X\omega_\alpha^2}{a^2 Z}, \qquad \alpha = 1, 2, \tag{8.19}$$

where

$$\begin{aligned}
X &= \Lambda C\Big(\varepsilon + \frac{\Lambda}{Va}\Big) + \frac{\Lambda\varepsilon}{Va},\\
Y &= \Lambda^2\Big(\varepsilon + \frac{\Lambda}{Va}\Big) + \Lambda^2 + R_s^2a^2,\\
Z &= \varepsilon + \Lambda C.
\end{aligned}$$

The values of ω_α^2 are found from the quadratic

$$\frac{\varepsilon^2C^2\Lambda}{(\varepsilon + \Lambda C)}\omega^4 - \Big(\frac{C\varepsilon Y}{Z} + \frac{X\Lambda}{Z} - B\Big)\omega^2 + \frac{Y\Lambda}{Z} - F = 0,$$

where

$$B = \frac{\Lambda^2 \varepsilon}{Va} + \Lambda^2 C + \varepsilon\Lambda + \frac{\Lambda^2}{Va} + R_s^2 a^2 C,$$
$$F = \Lambda^3 + \Lambda R_s^2 a^2.$$

We determine the minimum in a^2 of R^2 from (8.19), taking care to check that the appropriate value of ω_1^2 or ω_2^2 is positive. Before describing numerical results it is useful to first consider the equivalent problem without a salt field, but with the Vadasz inertia term present. In this case the analogous equations to (8.17) are

$$\begin{aligned} -\frac{\sigma}{Va}\Delta w &= \Delta w - R\Delta^* \theta, \\ \sigma\theta &= Rw - \Gamma, \\ C\sigma\Gamma &= -\Gamma - \Delta\theta. \end{aligned} \tag{8.20}$$

A normal mode analysis reveals that σ satisfies the equation

$$\frac{C\Lambda}{Va}\sigma^3 + \sigma^2\Lambda\Big(\frac{1}{Va} + C\Big) + \sigma\Big(\Lambda + \frac{\Lambda^2}{a^2} - R^2 a^2 C\Big) + \Lambda^2 - R^2 a^2 = 0.$$

One takes the real and imaginary parts of this equation to obtain

$$\omega^2 = \frac{\Lambda^2 - R^2 a^2}{\Lambda(C + 1/Va)},$$

and then

$$R_{osc}^2 = \frac{\Lambda}{a^2}\Big(\frac{1 + VaC}{VaC^2}\Big) + \frac{\Lambda^2}{a^2}\frac{1}{C^2 Va^2}. \tag{8.21}$$

The critical wave number from (8.21) is

$$a_c^2 = \pi^2\sqrt{1 + \frac{Va + Va^2 C}{\pi^2}}. \tag{8.22}$$

The oscillatory convection Rayleigh numbers are then found by inserting this into (8.21). It is worth observing that this is very different from the zero inertia situation of section 8.1. In that case the critical wavenumber for oscillatory convection is infinite over certain ranges. In the present situation the wavenumber is always finite for $0 < Va, C < \infty$. We find this also holds for the general double diffusive problem when $0 < Va, C, R_s^2 < \infty$.

Table 8.1 Critical values, $R^2_{osc} = Ra_{osc}$ and a^2_{osc}, for various values of C and Va, with $R^2_s = 1$, $\varepsilon = 10$.

Ra_{osc}	a^2_{osc}	C	Va
35.6932	551	0.03	1000
26.4334	634	0.04	1000
20.9779	708	0.05	1000
17.3841	775	0.06	1000
14.8389	838	0.07	1000
12.9423	897	0.08	1000
11.4746	952	0.09	1000
10.3051	1005	0.1	1000
34.5068	1095	0.03	2000
25.7144	1265	0.04	2000
20.4882	1416	0.05	2000
17.0253	1555	0.06	2000
14.5625	1684	0.07	2000
12.7255	1805	0.08	2000
11.2932	1919	0.09	2000
10.1529	2027	0.1	2000

8.2.1 Numerical results

We follow Straughan (2011b) and take $\varepsilon = 10$, which is a practical value. We fix $R^2_s = 1$ and report numerical results in tables 8.1 - 8.3. The stationary convection values for all C, Va are given by $R^2_{stat} = 1+4\pi^2 \approx 40.8696$ and $a^2_{stat} = \pi^2 \approx 9.8696$. In tables 8.1 - 8.3 we show values for oscillatory convection, Ra_{osc} and a^2_{osc}, when these are the dominant mode of convection. For all values in the tables when $C = 0, 0.01, 0.02$, stationary convection holds. There is a transition to oscillatory convection for C in the region $(0.02, 0.03)$. We consider values of $C = 0, 0.01, \ldots, 0.1$, and $Va = 1000, 2000, \ldots, 5000$. Other values have been computed, but those shown typify the behaviour witnessed on the surface in R^2, R^2_s, C space on the instability surface. Due to the very large range of critical values for the oscillatory wavenumber we find we only report values where the minimization is over integer values of a^2. The variation of critical wavenumbers in the general case is not unlike that predicted by equation (8.22) for the case where there is no salt field.

The pattern which emerges from tables 8.1 - 8.3 is that Ra_{osc} decreases as C increases for Va fixed, and Ra_{osc} decreases as Va increases for C fixed. As C increases the (negative) slope of the curve of Ra vs. C decreases appreciably. The critical wavenumber has a very strong variation. For C fixed, as Va increases a^2_{osc} increases appreciably. Likewise, for Va fixed, as C increases a^2_{osc} shows a strong increase.

Table 8.2 Critical values, $R^2_{osc} = Ra_{osc}$ and a^2_{osc}, for various values of C and Va, with $R_S^2 = 1$, $\varepsilon = 10$.

Ra_{osc}	a^2_{osc}	C	Va
34.1153	1642	0.03	3000
25.4766	1900	0.04	3000
20.3259	2132	0.05	3000
16.9062	2343	0.06	3000
14.4706	2539	0.07	3000
12.6479	2722	0.08	3000
11.2327	2895	0.09	3000
10.1021	3058	0.1	3000
33.9202	2191	0.03	4000
25.3578	2541	0.04	4000
20.2448	2852	0.05	4000
16.8466	3136	0.06	4000
14.4246	3399	0.07	4000
12.6111	3645	0.08	4000
11.2024	3875	0.09	4000
10.0767	4093	0.1	4000

Table 8.3 Critical values, $R^2_{osc} = Ra_{osc}$ and a^2_{osc}, for various values of C, with $R_S^2 = 1, \varepsilon = 10$.

Ra_{osc}	a^2_{osc}	C	Va
33.8032	2744	0.03	5000
25.2866	3184	0.04	5000
20.1960	3576	0.05	5000
16.8108	3933	0.06	5000
14.3969	4263	0.07	5000
12.5890	4570	0.08	5000
11.1842	4858	0.09	5000
10.0614	5131	0.1	5000

8.3 Soret effect

In this section we analyse a problem of convection in a Darcy porous material but we concentrate on incorporating a Soret effect. The Soret effect, discovered by Soret (1879), is one where mass may be separated by a temperature gradient, or alternatively, where the temperature gradient plays an important role on the diffusion flux. This effect is reviewed in depth by Platten (2006), and is investigated in convection studies in fluids and in fluid saturated porous media by Gaikwad and Kamble (2012), Capone et al. (2019), Gaikwad et al. (2009), Deepika (2018). An interesting article involving a negative Soret effect which plays an important role in density stratification in a mixture of fluids is that of Shevtsova et al. (2006). A further interesting article involving a mixture of different fluids with Soret coefficients of different signs is by García Fernández et al. (2022).

In the interests of clarity we introduce first the problem of double diffusion in a Darcy porous material when we incorporate the Soret effect, but no Cattaneo effects are present. We then analyse the analogous problem but when the diffusion flux satisfies a Cattaneo–like equation. The second scenario is very different from the former one and we highlight the novelties introduced by the combination of Cattaneo's law with a Soret effect.

8.3.1 Soret effect, Fick's law for concentration, Fourier for temperature

The governing equations for the evolution of the velocity, pressure, temperature and concentration fields in a fluid saturated Darcy porous material incorporating a Soret effect may be found in, for example, Straughan (2015a, page 41), and they are

$$
\begin{aligned}
&0=-\frac{1}{\rho_0}p_{,i}-\frac{\mu}{K\rho_0}v_i+\alpha T g k_i-\alpha_c g k_i C,\\
&v_{i,i}=0,\\
&\frac{1}{M}T_{,t}+v_iT_{,i}=\kappa\Delta T,\\
&\epsilon C_{,t}+v_iC_{,i}=\epsilon k_c\Delta C\pm\epsilon k_c S_T\Delta T,
\end{aligned}
\tag{8.23}
$$

where v_i, p, T and C are the velocity, pressure, temperature and concentration fields. The coefficients $\rho_0, \mu, K, \alpha, \alpha_c, g, \kappa, \epsilon, k_c, S_T$ are density, dynamic viscosity of the fluid, permeability, coefficient of thermal expansion of the fluid, coefficient of salt in the density, gravity, thermal diffusivity of the fluid, porosity, salt diffusivity, and the Soret coefficient. The coefficient M is defined after (8.25). In (8.23) the Soret effect is manifest by the term $\pm\epsilon k_c S_T\Delta T$, where at the outset of this section we allow for the possibility of a positive or negative Soret term.

We suppose the saturated porous material occupies the horizontal layer $\mathbb{R}^2\times\{z\in(0,d)\}$ with gravity acting in the negative z-direction, and the boundary conditions are

$$
C=C_L,\quad T=T_L,\quad z=0;\qquad C=C_U,\quad T=T_U,\quad z=d. \tag{8.24}
$$

The steady state solution with $\bar{v}_i\equiv 0$ corresponding to these boundary conditions is

$$
\bar{T}=-\beta z+T_L,\qquad \bar{C}=-\beta_c z+C_L, \tag{8.25}
$$

where

$$
\beta=\frac{T_L-T_U}{d},\qquad \beta_c=\frac{C_L-C_U}{d}.
$$

To analyse instability, perturbations u_i, π, θ, ϕ are introduced to this steady solution and the variables are non-dimensionalized with the scales

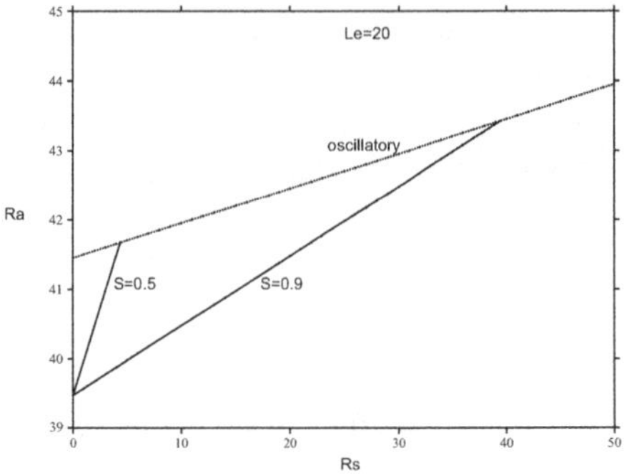

Fig. 8.1 Detail of transition behaviour between stationary convection and oscillatory convection. The transition occurs at $Ra = 41.673$, $Rs = 4.386$, when $S = 0.5$, and $Ra = 43.426$, $Rs = 39.478$, when $S = 0.9$.

$$
\begin{aligned}
&x_i = x_i^* d, \qquad t = t^* \mathcal{T}, \qquad u_i = u_i^* U, \qquad \theta = \theta^* T^\sharp, \qquad \phi = \phi^* C^\sharp, \\
&U = \frac{\kappa}{d}, \qquad \mathcal{T} = \frac{d^2}{M\kappa}, \qquad T^\sharp = \frac{\beta d^2}{\kappa}, \qquad S = \frac{S_T T^\sharp}{C^\sharp}, \qquad C^\sharp = \frac{\beta_c U d^2}{\epsilon k_c}, \\
&M = \frac{(\rho_0 c_p)_f}{(\rho_0 c)_m}, \qquad (\rho_0 c)_m = \epsilon(\rho_0 c_p)_f + (\rho_0 c)_s (1-\epsilon), \\
&Le = \frac{M\kappa}{k_c},
\end{aligned}
$$

where Le is the Lewis number and

$$
\kappa = \frac{k_m}{(\rho_0 c_p)_f}, \qquad k_m = k_s(1-\epsilon) + k_f \epsilon,
$$

a subscript f or s denotes fluid or solid, and a subscript m denotes the value in the porous medium.

With these scales and dropping stars, the non-dimensional perturbation equations become

$$
\begin{aligned}
&0 = -\pi_{,i} - u_i + Ra\theta k_i - Rs\phi k_i, \\
&u_{i,i} = 0, \\
&\theta_{,t} + u_i \theta_{,i} = H_1 w + \Delta\theta, \\
&Le\phi_{,t} + \frac{\kappa}{\epsilon k_c} u_i \phi_{,i} = w + \Delta\phi + H_2 S \Delta\theta,
\end{aligned}
\tag{8.26}
$$

where these equations hold on the domain $\mathbb{R}^2 \times \{z \in (0,1)\}$ for $t > 0$. The coefficients H_1 and H_2 are

$$
H_1 = \begin{cases} +1, & \text{when heated below,} \\ -1, & \text{when heated above,} \end{cases}
$$

and

$$
H_2 = \begin{cases} +1, & \text{for a positive Soret coefficient,} \\ -1, & \text{for a negative Soret coefficient.} \end{cases}
$$

The variable $w = u_3$ and Ra and Rs are the Rayleigh number and the salt Rayleigh number.

To analyse linear instability one discards the quadratic terms in (8.26) and then one seeks a time dependence like $e^{\sigma t}$, i.e. one puts $u_i = u_i(\mathbf{x})e^{\sigma t}$, with similar forms for π, θ, ϕ. This leads to the linear instability equations

$$\begin{aligned} 0 &= -\pi_{,i} - u_i + Ra\theta k_i - Rs\phi k_i, \\ u_{i,i} &= 0, \\ \sigma\theta &= H_1 w + \Delta\theta, \\ Le\sigma\phi &= w + \Delta\phi + H_2 S\Delta\theta. \end{aligned} \tag{8.27}$$

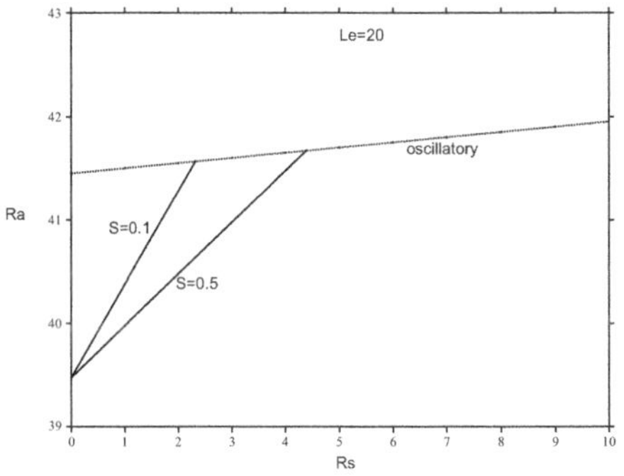

Fig. 8.2 Detail of transition behaviour between stationary convection and oscillatory convection. The transition occurs at $Ra = 41.673$, $Rs = 4.386$, when $S = 0.5$, and $Ra = 41.568$, $Rs = 2.322$, when $S = 0.1$.

From this point we concentrate on the case of heated below and positive Soret coefficient so that $H_1 = H_2 = +1$. By using a normal mode analysis one may find the Rayleigh numbers for stationary convection and for oscillatory convection. These may be shown to be

$$Ra_{stat} = Rs(1 - S) + \frac{\Lambda^2}{a^2},$$

and

$$Ra_{osc} = \frac{Rs}{Le} + \frac{(Le + 1)}{Le}\frac{\Lambda^2}{a^2}.$$

Minimizing these expressions in a^2 leads to

$$Ra_{stat} = 4\pi^2 + Rs(1 - S), \tag{8.28}$$

and

$$Ra_{osc} = 4\pi^2\frac{(Le + 1)}{Le} + \frac{Rs}{Le}, \tag{8.29}$$

with $a^2 = \pi^2$ in each case.

We take $Le = 20$ and show the behaviour of these curves in figures 8.1 and 8.2. The instability region lies above the stationary convection curve and the oscillatory convection curve. For example, in figure 8.1 the stationary convection curve $S = 0.9$

is as shown and this intersects the oscillatory convection curve at = 39.478. The steady solution is unstable if (Ra, Rs) is in the region above the curve marked $S = 0.9$ or above the oscillatory convection curve if $Rs \geq 39.478$.

8.3.2 Soret effect, Cattaneo's law for concentration, Fourier for temperature

We now modify equations (8.23) to allow for a Cattaneo effect in the equation for the concentration C. Thus, we now propose instead of (8.23) the system of partial differential equations

$$
\begin{aligned}
0 &= -\frac{1}{\rho_0}p_{,i} - \frac{\mu}{K\rho_0}v_i + \alpha T g k_i - \alpha_c g k_i C, \\
v_{i,i} &= 0, \\
\frac{1}{M}T_{,t} + v_i T_{,i} &= \kappa \Delta T, \\
\epsilon C_{,t} + v_i C_{,i} &= -\epsilon J_{i,i}\,, \\
\tau(J_{i,t} + v_i J_{,i}) + J_i &= -k_c C_{,i} \mp k_c S_T T_{,i},
\end{aligned}
\tag{8.30}
$$

where $\mathbf{J}$ is the diffusion flux and $\tau > 0$ is a relaxation time. Observe that when $\tau = 0$ equations (8.30) reduce to (8.23).

The steady state solution (8.30) is again $\bar{v}_i \equiv 0$ with (8.25) and now

$$\bar{J}_3 = k_c\beta_c \pm k_c S_T \beta.$$

To analyse stability of this steady solution we introduce perturbations u_i, π, θ, ϕ and j_i and non-dimensionalize as in section 8.3.1 employing (3.15) for the non-dimensional version of τ. We concentrate on heating from below and a positive Soret coefficient and one may arrive at the following system of non-dimensional perturbation equations

$$
\begin{aligned}
0 &= -\pi_{,i} - u_i + Ra\theta k_i - Rs\phi k_i, \\
u_{i,i} &= 0, \\
\theta_{,t} + u_i\theta_{,i} &= w + \Delta\theta, \\
Le\phi_{,t} + \frac{\kappa}{\epsilon k_c}u_i\phi_{,i} &= w - j_{i,i}\,, \\
Sg(j_{i,t} + u_k j_{i,k}) + j_i &= -\phi_{,i} - S\theta_{,i}\,,
\end{aligned}
\tag{8.31}
$$

where j_i is the non-dimensional perturbation to J_i.

To analyse linear instability one discards the quadratic terms in (8.31) and then one seeks a time dependence like $e^{\sigma t}$, i.e. one puts $u_i = u_i(\mathbf{x})e^{\sigma t}$, with similar forms for π, θ, ϕ, j_i. After removing π this leads to the linear instability equations

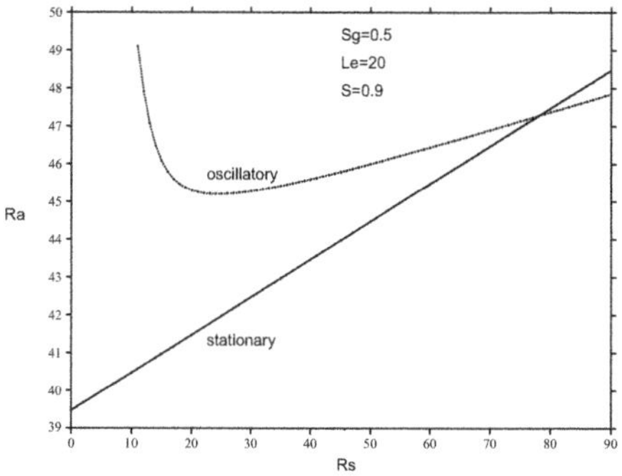

Fig. 8.3 Detail of transition behaviour between stationary convection and oscillatory convection. The transition occurs at $Ra = 47.283$, $Rs = 78.05$.

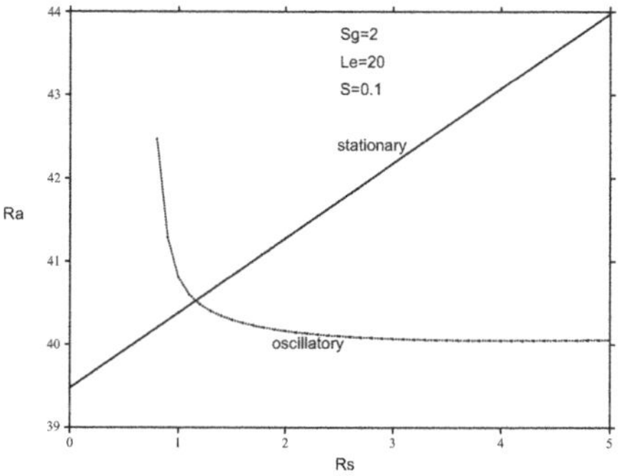

Fig. 8.4 Detail of transition behaviour between stationary convection and oscillatory convection. The transition occurs at $Ra = 47.474$, $Rs = 1.15$.

$$
\begin{aligned}
&0 = -Ra\Delta^*\theta + Rs\Delta^*\phi + \Delta w,\\
&\sigma\theta = w + \Delta\theta,\\
&Le\sigma\phi = w - \Gamma,\\
&(Sg\sigma + 1)\Gamma = -\Delta\phi - S\Delta\theta,
\end{aligned}
\tag{8.32}
$$

where $\Gamma = j_{i,i}$. Solution of this system of equations leads to the following cubic equation in σ,

$$
\begin{aligned}
\Lambda(\sigma + \Lambda)(\sigma^2 SgLe + \sigma Le + \Lambda) - RsS\Lambda a^2&\\
+ (\sigma Sg + 1)(\sigma + \Lambda)Rsa^2 = Raa^2(Sg\sigma^2 Le + \sigma Le + \Lambda).&
\end{aligned}
\tag{8.33}
$$

To find the stationary convection boundary one puts $\sigma = 0$ in (8.33) whereas to derive the oscillatory convection threshold one chooses $\sigma = i\omega$. After taking the real and imaginary parts of the resulting equation one obtains a quadratic equation for Raa^2 of form

$$
(Raa^2)^2 - Raa^2 B + C = 0,
$$

where

$$
B = 2Le\Lambda^2 + \frac{Le}{Sg}\Lambda + Rsa^2(2 + \Lambda Sg),
$$

and

Table 8.4 Critical values, Ra_{stat}, $R^2_{osc} = Ra_{osc}$ and a^2_{osc}, vs. Rs. $Le = 20$, $S = 0.1$ $Sg = 0.5$.

Ra_{stat}	Ra_{osc}	a^2_{osc}	Rs
42.1784	59.2240	33.6	3.0
42.2684	56.1599	30.2	3.1
42.3584	53.5263	27.3	3.2
42.4484	51.2502	24.6	3.3
42.5384	49.2027	22.1	3.4
42.6284	47.4945	20.0	3.5
42.7184	46.0465	18.0	3.6
42.8084	44.8587	16.2	3.7
42.8984	43.9167	14.5	3.8
42.9884	43.2197	13.1	3.9
43.0784	**42.7589**	11.8	4.0

Table 8.5 Critical values, Ra_{stat}, $R^2_{osc} = Ra_{osc}$ and a^2_{osc}, vs. Rs. $Le = 20$, $S = 0.5$ $Sg = 0.5$.

Ra_{stat}	Ra_{osc}	a^2_{osc}	Rs
43.4784	43.9666	12.5	8.0
43.5284	43.8855	12.3	8.1
43.5784	43.8128	12.1	8.2
43.6284	43.7476	12.0	8.3
43.6784	43.6888	11.8	8.4
43.7284	**43.6359**	11.7	8.5
43.7784	43.5882	11.6	8.6
43.8284	43.5450	11.5	8.7
43.8784	43.5059	11.4	8.8
43.9284	43.4703	11.4	8.9
43.9784	43.4374	11.3	9.0

$$C = \Big[Le\Big(\Lambda^2 + \frac{\Lambda}{Sg}\Big) + Rsa^2\Big]\Big[\Lambda^2\Big(\frac{1+Le}{Le}\Big) + Rsa^2\frac{(1+\Lambda Sg)}{Le}\Big] - \Lambda^4 - \Lambda^2 Rsa^2(1-S).$$

Then the oscillatory convection boundary is found by *carefully* minimizing

$$Ra = \frac{B \pm \sqrt{B^2 - 4LeC}}{2a^2Le},$$

numerically in a^2. The stationary convection boundary follows from

$$Ra_{stat} = \frac{\Lambda^2}{a^2} + Rs(1-S).$$

Hence, minimizing in a^2

$$Ra_{stat} = 4\pi^2 + Rs(1-S), \tag{8.34}$$

Table 8.6 Critical values, Ra_{stat}, $R^2_{osc} = Ra_{osc}$, a^2_{osc}, at the transition from stationary to oscillatory convection. $Le = 20$.

Ra_{stat}	Ra_{osc}	a^2_{osc}	Rs	S	Sg
41.4584	41.6313	11.6	2.2	0.1	1
41.5484	41.4344	11.1	2.2	0.1	1
40.4684	40.5981	10.6	1.1	0.1	2
40.5584	40.4807	10.4	1.2	0.1	2
41.6284	41.6703	10.7	4.3	0.5	1
41.6784	41.6422	10.7	4.4	0.5	1
40.5784	40.5793	10.3	2.2	0.5	2
40.6284	40.5603	10.2	2.3	0.5	2
47.2784	47.2820	10.9	78.0	0.9	0.5
47.2884	47.2867	10.9	78.1	0.9	0.5
43.4184	43.4234	10.4	39.4	0.9	1
43.4284	43.4281	10.4	39.5	0.9	1
41.4584	41.4622	10.1	19.9	0.9	2
41.4684	41.4669	10.1	20.0	0.9	2

with $a^2_{stat} = \pi^2$.

The stationary convection threshold is, as expected, the same as in section 8.3.1.

While the stationary convection boundary is the same the oscillatory convection one is very different from section 8.3.1. We provide numerical details in figures 8.3 and 8.4 where $Le = 20$ and $Sg = 0.5, S = 0.9$ and $Sg = 2, S = 0.1$. It is seen that the oscillatory convection curve is very nonlinear reaching a minimum in Rs and then increasing. The curve behaviour and the intersection with the stationary convection curve is strongly dependent on the parameters S and Sg. We have concentrated on values of $S = 0.1, 0.5$ and 0.9 and have selected Sg in the range $(0, 2]$ with $Le = 20$. The qualitative behaviour of the oscillatory convection curves are all not dissimilar, but the quantitative values vary a lot. Tables 8.4, 8.5 and 8.6 show how the transition from stationary to oscillatory convection varies in Rs as S and Sg change. These tables also show how the oscillatory convection wavenumber varies with Rs.

The instability boundary in figures 8.3 and 8.4 is the stationary convection curve until the transition point and then the oscillatory convection curve.

It is quite clear that when a Cattaneo theory is employed the behaviour of the critical Rayleigh and wave numbers changes strongly with variations in the Soret coefficient.

8.4 Cattaneo double diffusion in a fluid

A thorough analysis of the double diffusion problem (including temperature, T, and concentration, S,) in a linearly viscous incompressible fluid is given by Hughes et al. (2021). These writers employ a Cattaneo equation for both the heat flux and for the diffusion flux. Their basic steady solution is one where T and S are linear functions of z, the velocity is zero, and they employ a Boussinesq approximation. They denote the heat flux by $\mathbf{q}_T$, the salt flux by $\mathbf{q}_S$, and propose a Cattaneo–Christov theory for both fluxes. These writers work directly with the divergence of both $\mathbf{q}_T$ and $\mathbf{q}_S$ which they denote as Q_T and Q_S. This then leads to the following system of nonlinear, non-dimensional equations

$$
\begin{aligned}
&\frac{1}{Pr}\frac{Du_i}{Dt} = -\frac{\partial p}{\partial x_i} + RaTk_i - RsSk_i + \Delta u_i,\\
&\frac{\partial u_i}{\partial x_i} = 0,\\
&\frac{DT}{Dt} = w - Q_T,\\
&2C_T\frac{DQ_T}{Dt} = -Q_T - \Delta T,\\
&\frac{DS}{Dt} = w - Q_S,\\
&2C_S\frac{DQ_S}{Dt} = -Q_S - \frac{\kappa_S}{\kappa}\Delta S,
\end{aligned}
$$

where D/Dt is the material derivative, Ra, Rs are the Rayleigh and salt Rayleigh numbers, C_T and C_S are Cattaneo numbers for the heat flux and salt flux, and Pr, κ_S, κ are the Prandtl number, and the salt and thermal diffusivities.

It is worth pointing out that in the case of heating and salting below, the model of Hughes et al. (2021) has three skew-symmetric terms in the linear operator, each of which could affect oscillatory instability. Hence, one might expect a rich structure for the transition to convection surface.

Hughes et al. (2021) develop a complete linear instability analysis for stress free boundary conditions. They derive a fifth order equation for the growth rate σ. In their analysis Rs and Ra may be positive or negative so they encompass all cases. They identify specific ranges of Rs as a function of C_T^{-n} with distinct regimes emerging for $n \in (0,1]$, $n \in (1,2)$, $n = 2$ and $n > 2$. The bifurcation picture is complicated but an intricate analysis is performed exactly. The oscillatory convection Rayleigh number is shown to have a bimodal behaviour in the square of the wavenumber which depends critically on the ratio of κ_S/κ. Their analysis is very detailed and they note that the Cattaneo theories lead to a much richer problem than the analogous Fourier one. They further point out that the dependence of the critical Rayleigh number upon the wavenumber is decidedly non trivial. The paper of Hughes et al. (2021) is certainly a very influential one in the theory of convection involving Cattaneo flux equations.

Chapter 9
Cattaneo convection with rotation or a magnetic field

9.1 Magnetohydrodynamic convection with a Cattaneo law

The problem of thermal convection in a layer of fluid with a superimposed magnetic field is developed extensively in Chandrasekhar (1981, chapter 3). The work described in Chandrasekhar (1981) employs Fourier's law to describe the relation between the heat flux and the temperature gradient. The inclusion of a Cattaneo effect into the analysis of thermal convection in MHD was made by Bissell (2016). He assumes the fluid occupies the infinite layer $\mathbb{R}^2 \times \{z \in (0,d)\}$ and the magnetic field is in the direction of gravity acting vertically.

The basic equations are those of Chandrasekhar (1981) except that one includes an appropriate time derivative for the heat flux in order to encompass the Cattaneo theory of heat flow. Bissell (2016) employs CCJ theory and so his equations are

$$\begin{aligned}
&v_{i,t} + v_j v_{i,j} = \frac{\mu}{\rho_0} H_j H_{i,j} - p_{,i} + \nu \Delta v_i + g\alpha k_i T,\\
&v_{i,i} = 0,\\
&\rho c_v (T_{,t} + v_i T_{,i}) = -Q_{i,i}\,,\\
&\tau(Q_{i,t} + v_j Q_{i,j}) = -Q_i - kT_{,i}\,,\\
&H_{i,t} + v_j H_{i,j} = H_j v_{i,j} + \eta \Delta H_i\,,\\
&H_{i,i} = 0.
\end{aligned} \tag{9.1}$$

In equations (9.1) the variables v_i, T, Q_i are as in other chapters, H_i denotes the magnetic field, p is the pressure which incorporates a magnetic term, cf. Chandrasekhar (1981, page 148), and η is the constant magnetic resistivity.

The basic solution studied by Bissell (2016) is

$$\begin{aligned}
&\bar{v}_i \equiv 0, \qquad \bar{T} = -\beta z + T_L,\\
&\bar{\mathbf{Q}} = (0,0,k\beta), \qquad \bar{H}_i = \bar{H}\delta_{i3},
\end{aligned}$$

B. Straughan, *Thermal Convection with a Cattaneo Flux Law*, Advances in Mechanics and Mathematics 53, https://doi.org/10.1007/978-3-031-80885-2_9

where $\beta = (T_L - T_U)/d$, $\bar{H}$ is a constant and T_L, T_U are the constant lower and upper temperatures of the layer with $T_L > T_U$.

Perturbations to (9.1) are introduced as u_i, θ, q_i, π and h_i and then Bissell (2016) finds the linearized, non-dimensional perturbation equations are

$$\begin{aligned}
\frac{\partial u_i}{\partial t} &= -\frac{\partial \pi}{\partial x_i} + \Delta u_i + R\theta k_i + \underline{\frac{Q}{Pm}\frac{\partial h_i}{\partial z}},\\
\frac{\partial u_i}{\partial x_i} &= 0,\\
Pr\frac{\partial \theta}{\partial t} &= Rw - \underline{\underline{\frac{\partial q_i}{\partial x_i}}},\\
Sg\frac{\partial q_i}{\partial t} &= -q_i - \underline{\underline{\frac{\partial \theta}{\partial x_i}}},\\
Pm\frac{\partial h_i}{\partial t} &= \underline{Pm\frac{\partial u_i}{\partial z}} + \Delta h_i,\\
\frac{\partial h_i}{\partial x_i} &= 0,
\end{aligned} \tag{9.2}$$

where Pr is the Prandtl number, and $Pm = \nu/\eta$ is the magnetic Prandtl number. These equations hold on the horizontal layer $\mathbb{R}^2 \times \{z \in (0,1)\}$ for $t > 0$. The functions u_i, θ, q_i, π and h_i satisfy a condition of periodicity in x, y and also satisfy the boundary conditions

$$u_i = 0, \quad \theta = 0, \quad h_i = 0,$$

for fixed surfaces with u_i satisfying stress free boundary conditions for two stress free surfaces.

In classical MHD with Fourier's law ($Sg = 0$) the linear operator on the right of (9.2) is not symmetric. The terms $(Q/Pm)\,\partial h_i/\partial z$ in $(9.2)_1$ and $Pm\,\partial u_i/\partial z$ in $(9.2)_5$, with one underline, represent a skew-symmetric part to the linear operator when written as

$$AU_t = LU,$$

where A is a diagonal matrix and L effectively represents the right hand side of (9.2), with $U = (u_1, u_2, w, q_1, q_2, q_3, h_1, h_2, h_3)^T$. The presence of this skew-symmetric part leads to oscillatory instability, Chandrasekhar (1981, pages 181–186). However, there is a second skew-symmetric part in L from the terms $-\partial q_i/\partial x_i$ in $(9.2)_3$ and $-\partial\theta/\partial x_i$ in $(9.2)_4$ (the doubly underlined terms). This skew - symmetric part has been seen to lead to oscillatory instability in section 3.3. Hence, for the magnetic Bénard problem with a Cattaneo heat flux law there are two independent skew - symmetric effects and one may expect a richer structure in the instability thresholds (the bifurcation diagram).

Bissell (2016) employs a normal mode analysis to analyse the instability thresholds of (9.2) in detail, for two stress free surfaces. He also deals with fixed surfaces by employing a Chebyshev tau numerical method. He mainly deals with the situation where $Pm = 0$. Many detailed results are presented by Bissell (2016) and we refer to that paper for a full analysis. One important fact observed by Bissell (2016, page 16) is, *...for sufficiently large Chandrasekhar numbers Q, oscillatory convection can represent the preferred form of instability even when the Cattaneo* effect *is small.*

Eltayeb et al. (2020) continue this analysis with much detail when $Pm > 0$ and they employ the boundary condition $\partial h_3/\partial z = 0$. They decompose u_i and h_i into their poloidal and solenoidal components and perform an intricate bifurcation analysis of the transition to convection. They discover very interesting effects when $Q >> 1$ and find asymptotic relations between Q and the size of the Cattaneo relaxation effect.

9.2 Maxwell's displacement current

If the electrical permittivity, ϵ, and the magnetic permeability, μ, are constant, then Maxwell's equations are comprised of Faraday's law, (1.41),

$$\mu \frac{\partial \mathbf{H}}{\partial t} = -curl\, \mathbf{E}, \tag{9.3}$$

and Amperé's law with Maxwell's displacement correction, (1.42),

$$\epsilon \frac{\partial \mathbf{E}}{\partial t} + \sigma \mathbf{E} = curl\, \mathbf{H}, \tag{9.4}$$

where $\mathbf{E}$ and $\mathbf{H}$ are the electric and magnetic fields and where we have replaced the current density $\mathbf{j}$ by using Ohm's law $\mathbf{j} = \sigma \mathbf{E}$. The magnetic induction, $\mathbf{B}$, and the electric displacement, $\mathbf{D}$, are given by $\mathbf{B} = \mu \mathbf{H}$, and $\mathbf{D} = \epsilon \mathbf{E}$. These equations with further memory effects are examined by Graffi (1936, 1963, 1965), see also Franchi and Straughan (2023, pages 18–22).

As observed in section 1.3.1 there is an analogy between equations (9.3) and (9.4) and those of Cattaneo in heat flow, namely

$$\begin{aligned} \rho c_p \frac{\partial T}{\partial t} &= -\frac{\partial q_i}{\partial x_i}, \\ \tau \frac{\partial q_i}{\partial t} + q_i &= -k \frac{\partial T}{\partial x_i}. \end{aligned} \tag{9.5}$$

If we replace T with $\mathbf{H}$ and q_i with $\mathbf{E}$ then system (9.5) is formally like (9.3) and (9.4). This is clear if we write each system of equations in the following forms

$$\begin{pmatrix} \mu & 0 \\ 0 & \epsilon \end{pmatrix} \frac{\partial}{\partial t} \begin{pmatrix} \mathbf{H} \\ \mathbf{E} \end{pmatrix} = - \begin{pmatrix} 0 & 0 \\ 0 & \sigma \end{pmatrix} \begin{pmatrix} \mathbf{H} \\ \mathbf{E} \end{pmatrix} + \begin{pmatrix} 0 & -curl \\ curl & 0 \end{pmatrix} \begin{pmatrix} \mathbf{H} \\ \mathbf{E} \end{pmatrix}$$

and

$$\begin{pmatrix} \rho c_p & 0 \\ 0 & \tau \end{pmatrix} \frac{\partial}{\partial t} \begin{pmatrix} T \\ \mathbf{q} \end{pmatrix} = - \begin{pmatrix} 0 & 0 \\ 0 & -1 \end{pmatrix} \begin{pmatrix} T \\ \mathbf{q} \end{pmatrix} + \begin{pmatrix} 0 & -\frac{\partial}{\partial \mathbf{x}} \\ -k\frac{\partial}{\partial \mathbf{x}} & 0 \end{pmatrix} \begin{pmatrix} T \\ \mathbf{q} \end{pmatrix}$$

The linear operators on the right in each case contain a skew-symmetric part, namely

$$\begin{pmatrix} 0 & -curl \\ curl & 0 \end{pmatrix}$$

and

$$\begin{pmatrix} 0 & -\frac{\partial}{\partial \mathbf{x}} \\ -k\frac{\partial}{\partial \mathbf{x}} & 0 \end{pmatrix}$$

They are skew-symmetric under the correct boundary conditions, cf. Chandrasekhar (1981), Cimatti (2024). We have seen in chapter 3 that the skew - symmetric operator in the Cattaneo system of equations may lead to oscillatory instability in thermal convection. It is interesting to ask the same question in magnetohydrodynamic convection if Maxwell's displacement current is included. We comment on this below.

Maxwell's displacement current has been included in MHD by Stupelis (2000, 2004a,b, 2007). Stupelis (2004a) does suggest a model for MHD which incorporates temperature effects. His system may be written (with our notation) as

$$\begin{aligned}
&\frac{\partial v_i}{\partial t} + v_j\frac{\partial v_i}{\partial x_j} = -\frac{1}{\rho}\frac{\partial p}{\partial x_i} - \nu\Delta v_i + \alpha g T k_i + \frac{\mu}{\rho}(\mathbf{j}\times\mathbf{H})_i\,,\\
&\frac{\partial v_i}{\partial x_i} = 0,\\
&\frac{\partial T}{\partial t} + v_i\frac{\partial T}{\partial x_i} = \kappa\Delta T,\\
&\epsilon\frac{\partial \mathbf{E}}{\partial t} - curl\,\mathbf{H} + \mathbf{j} = \mathbf{j}^{(e)},\\
&\mu\frac{\partial \mathbf{H}}{\partial t} + curl\,\mathbf{E} = 0,\\
&\frac{\partial H_i}{\partial x_i} = 0,
\end{aligned}\tag{9.6}$$

where we treat $\epsilon, \mu, \nu, \kappa$ as constants. The function $\mathbf{j}^{(e)}$ is a density vector arising from the action of external electromotive forces. The current $\mathbf{j}$ is given by

$$\mathbf{j} = \sigma\big[\mathbf{E} + \mu(\mathbf{v}\times\mathbf{H})\big]\,,\tag{9.7}$$

where the conductivity, σ, is constant. Stupelis (2004a) analyses existence of a solution to (9.6). I am not aware of any analysis of thermal convection with this model, but it may be an interesting problem given that the equation for the electric field is somewhat similar to the Cattaneo equation for the heat flux. Such a problem could present two sources of oscillatory convection, namely that in classical MHD, cf. Chandrasekhar (1981), and that due to Maxwell's displacement current. Of course, if in addition the equation for the heat flux was an evolution equation of Cattaneo type then there would be three potential sources of oscillatory convection which could lead to a complicated bifurcation behaviour.

9.3 Cattaneo convection with rotation

The problem of thermal convection in a horizontal layer of fluid which is rotating about a vertical axis in line with gravity and incorporating the Cattaneo law for the heat flux is studied in some detail by Hughes et al. (2022). These writers work directly with the divergence of the heat flux, i.e. $\Gamma = q_{i,i}$, and adopt a Cattaneo - Christov theory. Thus, in terms of the material derivative $D/Dt = \partial/\partial t + \mathbf{v}.\nabla$ their non-dimensional governing equations are

$$
\begin{aligned}
&\frac{1}{Pr}\frac{D\mathbf{u}}{Dt} + T\,\mathbf{k}\times\mathbf{u} = -\nabla p + Ra\theta\mathbf{k} + \Delta\mathbf{u},\\
&div\,\mathbf{u} = 0,\\
&\frac{D\theta}{Dt} = w - \Gamma,\\
&C\frac{D\Gamma}{Dt} = -\Gamma - \Delta\theta,
\end{aligned}
$$

where Pr is the Prandtl number, $C = \tau\kappa/d^2$ is a Cattaneo number, u_i and θ are the velocity and temperature perturbations, and $Ta = T^2$ is the Taylor number which measures the rotation rate of the layer.

Hughes et al. (2022) employ stress free boundary conditions and develop a linear instability analysis. They write the velocity in terms of its poloidal and toroidal components, $\mathcal{P}, \mathcal{T}$, as

$$
\mathbf{u} = \nabla\times(\nabla\times\mathcal{P}\mathbf{k}) + \nabla\times\mathcal{T}\mathbf{k}.
$$

This then leads them to solve four partial differential equations

$$
\begin{aligned}
&\frac{1}{Pr}\frac{\partial\mathcal{T}}{\partial t} - T\frac{\partial\mathcal{P}}{\partial z} = \Delta\mathcal{T},\\
&\frac{1}{Pr}\frac{\partial\Delta\mathcal{P}}{\partial t} + T\frac{\partial\mathcal{T}}{\partial z} = -Ra\theta + \Delta^2\mathcal{P},\\
&\frac{\partial\theta}{\partial t} = -\Delta^*\mathcal{P} - \Gamma,\\
&C\frac{\partial\Gamma}{\partial t} = -\Gamma - \Delta\theta.
\end{aligned}
$$

The work of Hughes et al. (2022) is technical but very interesting. They show that the Cattaneo effect modifies the behaviour markedly compared to classical rotating Bénard convection. There is a strong relation between the parameters $T^{1/3}, C$ and Pr which is analysed in detail.

For technical details we refer to Hughes et al. (2022). It is worth pointing out that the results are succinctly described in their conclusion. They deduce that the Cattaneo effect may be important in the Earth's outer core. These writers also point out that the modes in their analysis may have large critical horizontal wave numbers which can have an effect on the vertical structure of convective motion.

Chapter 10
LTNE convection with Cattaneo effect in the solid

10.1 Darcy theory, LTNE

Local thermal non-equilibrium theory in a porous medium is a progressively important area of modern research. This area is where the fluid in the pores may have a different temperature to the temperature of the solid skeleton making up the porous medium. It is usual to use the notation LTNE to denote local thermal non-equilibrium. A significant development in understanding flow in an LTNE environment was provided by Banu and Rees (2002). An excellent account of the historical development of the problem of thermal convection in a porous layer under an LTNE environment is provided by Bidin and Rees (2020), with Rees (2009) adding references to where the topic goes back to 1926. Banu and Rees (2002) presented a rigorous, accurate, linearized instability analysis of LTNE thermal convection in a porous horizontal layer. This analysis was corroborated by Straughan (2006) who employed energy stability theory to show that the work of Banu and Rees (2002) continued to be true also for a fully nonlinear stability analysis. Following the seminal article of Banu and Rees (2002), many papers followed in the LTNE vein. A particularly important contribution is by Rees et al. (2008) who demonstrated that local thermal non-equilibrium theory leads to very different results to the case where there is only one temperature, and indeed, in a certain limit it can appear to be hyperbolic. Other important analyses are by Rees and Bassom (2010), Rees and Pop (2010), and Rees (2003), who further confirmed that the different pore fluid and solid temperatures may yield significant changes when compared with single temperature theory. Important recent experimental work on LTNE effects on heat transport in porous materials, see Gossler et al. (2019), Baek et al. (2022), strongly confirms the need for the fundamental analyses of Rees and his co-workers. Following the instability analysis of Banu and Rees (2002) there has been a huge interest in this field, as witnessed by for example, Bidin and Rees (2020), Capone and Gentile (2018), Freitas et al. (2022), Lakshmi et al. (2022), Siddheshwar et al. (2021), Nandal et al. (2023), Shivakumara et al. (2023), Shivakumara and Raghunatha (2022), Shyabal et al. (2023), Straughan (2015a), and the references cited there. Articles worth highlighting are those of

B. Straughan, *Thermal Convection with a Cattaneo Flux Law*, Advances in Mechanics and Mathematics 53, https://doi.org/10.1007/978-3-031-80885-2_10

Rees (2009, 2010, 2011). These references provide important information on how to calculate the heat transfer coefficient between the solid and fluid temperatures. This coefficient is usually denoted by H and leads to a non-dimensional parameter we call the Rees number. Interest in this field is driven by the many real life applications such as extracting water from an aquifer, Böttcher et al. (2019), Gossler et al. (2019), Baek et al. (2022), designing hydrogen fuel cells, Damm and Fedorov (2006), producing tube refrigerators for space applications, Ashwin et al. (2010), in renewable engineering such as electricity creation from a solar pond, Mahfoudh et al. (2019), in hydraulic stimulation of enhanced geothermal systems, Fan et al. (2024), in nuclear engineering design, the behaviour of internal combustion engines, or producing fibre-reinforced materials, as is described in Nandal et al. (2023) and many more examples can be found in the book by Straughan (2015a, pages 4,5).

In the context of the Darcy theory for a porous medium the appropriate equations of Banu and Rees (2002) may be taken to be

$$
\begin{aligned}
&v_i = -\frac{K}{\mu}\frac{\partial p}{\partial x_i} + \frac{\rho_f g \alpha K}{\mu} T_f k_i ,\\
&\frac{\partial v_i}{\partial x_i} = 0,\\
&\phi(\rho c)_f \frac{\partial T^f}{\partial t} + (\rho c)_f\, v_i \frac{\partial T^f}{\partial x_i} = \phi k_f \Delta T^f + h(T_s - T_f),\\
&(1-\phi)(\rho c)_s \frac{\partial T^s}{\partial t} = (1-\phi) k_s \Delta T^s - h(T^s - T^f).
\end{aligned}
\tag{10.1}
$$

In these equations v_i, T, p denote the pore averaged velocity, the temperature, and the pressure. A sub or superscript f or s indicates whether the quantity refers to the fluid or solid component. The constants $\phi, K, \mu, \rho, g, \alpha, k, h$ are the porosity (fluid volume divided by total volume of fluid and solid), the permeability, the dynamic viscosity of the fluid, the density, gravity, the coefficient of expansion of the fluid, the thermal conductivity, and an interaction coefficient describing how efficiently heat is transferred from the solid to fluid or vice-versa. We stress that the f and s refer to quantities associated to the fluid or solid, respectively, e.g. k_f or equivalently k^f mean the thermal conductivity of the fluid.

We now consider a layer of porous material saturated with fluid and contained between the planes $z = 0$ and $z = d$. To consider the problem of thermal convection we suppose the temperatures of the solid, T_s, and of the fluid, T_f, are maintained as constants on the planes $z = 0$ and $z = d$, such that

$$
T_s = T_f = T_L , \quad \text{on} \quad z = 0; \qquad T_s = T_f = T_U , \quad \text{on} \quad z = d; \tag{10.2}
$$

where T_L, T_U are constants such that $T_L > T_U$.

The steady solution to (10.1) and (10.2) whose stability is of interest is

$$
\bar{v}_i \equiv 0, \qquad \bar{T}_f = \bar{T}_s = -\beta z + T_L . \tag{10.3}
$$

In an investigation of instability or stability of the steady solution (10.3) one introduces perturbations $u_i, \theta, \varphi, \pi$ as follows

$$v_i = \bar{v} + u_i, \quad p = \bar{p} + \pi, \quad T_f = \bar{T}_f + \theta_f, \quad T_s = \bar{T}_s + \theta_s .$$

The perturbation equations which arise from (10.1) are non-dimensionalized with the velocity, pressure, temperature, time and length scales of

$$U = \frac{\phi k_f}{(\rho c)_f d}, \quad P = \frac{\mu d U}{K}, \quad T^\sharp = Ud\sqrt{\frac{\mu \beta c_f}{\phi k_f g \alpha K}}, \quad \mathcal{T} = \frac{(\rho c)_f d^2}{k_f}, \quad L = d.$$

Important parameters are the Rayleigh number, Ra, given by

$$Ra = R^2 = d^2 \rho_f^2 \sqrt{\frac{\beta c_f g \alpha K}{\phi k_f \mu}},$$

and the Rees number, H, given by

$$H = \frac{hd^2}{\epsilon k_f}.$$

We also need the non-dimensional parameters

$$\gamma = \frac{\phi k_f}{(1-\phi)k_s}, \qquad A = \frac{\rho_s c_s k_f}{\rho_f c_f k_s}.$$

The parameters H and γ were introduced by Banu and Rees (2002). With θ and φ being the non-dimensional versions of θ_f and θ_s, the non-dimensional perturbation equations which result in this way are

$$\begin{aligned} u_i &= -\frac{\partial \pi}{\partial x_i} + R\theta k_i , \\ \frac{\partial u_i}{\partial x_i} &= 0, \\ \frac{\partial \theta}{\partial t} + u_i \frac{\partial \theta}{\partial x_i} &= Rw + \Delta\theta + H(\varphi - \theta), \\ A\frac{\partial \varphi}{\partial t} &= \Delta\varphi - H\gamma(\varphi - \theta). \end{aligned} \tag{10.4}$$

The boundary conditions to be satisfied are

$$w = 0, \quad \theta = 0, \quad \varphi = 0, \qquad \text{on} \quad z = 0, 1, \tag{10.5}$$

together with $u_i, \pi, \theta, \varphi$ satisfying a plane tiling periodicity in x, y.

The solution to equations (10.4), (10.5) is discussed at length by Banu and Rees (2002). This problem together with many others related to LTNE convection with a

Fourier law of heat conduction is discussed in much detail in the book by Straughan (2015a). Since the current book is about thermal convection with a Cattaneo heat flux law, we now turn to the porous convection LTNE problem but incorporating a Cattaneo law.

10.1.1 Darcy theory, LTNE, Cattaneo in solid

Straughan (2013) further analysed the Banu and Rees (2002) instability problem but he allowed the heat flux in the solid skeleton to be governed by a Cattaneo flux law rather than a Fourier one. This work has been developed in various directions by, for example, Eltayeb (2015), Shivakumara et al. (2015a,b), Eltayeb and Elbashir (2017), Hema et al. (2020, 2021), Mamatha et al. (2022), Capone and Gianfrani (2022). One of the reasons for the interest in this field is the many potential application areas it applies to. A potentially important application is to magma flow in a volcano, see e.g. Vieira et al. (2021), Allocca et al. (2022), Toy et al. (2019), Singh et al. (2022), Bagdassarov and Fradkov (1993), De Campos et al. (2005). Given the recent seismic and volcanic activity in Iceland and in the Campi Flegrei region near Pozzuoli, an understanding of the impact on the environment is of vital importance. Another possible application with huge consequences for the environment is to renewable energy such as electricity generation via a solar pond, see e.g. Dineshkumar and Raja (2022), Mahfoudh et al. (2019), Wang et al. (2018).

Straughan (2013) argues that ... *because the second sound effect appears greater in solids, especially those involved in porous metallic foams, we restrict attention to this effect in solids, and retain the usual Fourier heat transfer law in the fluid.* The basic equations written down by Straughan (2013) are

$$
\begin{aligned}
&v_i = -\frac{K}{\mu}\frac{\partial p}{\partial x_i} + \frac{\rho_f g \alpha K}{\mu} T_f k_i \, , \\
&\frac{\partial v_i}{\partial x_i} = 0, \\
&\phi(\rho c)_f \frac{\partial T^f}{\partial t} + (\rho c)_f \, v_i \frac{\partial T^f}{\partial x_i} = \phi k_f \Delta T^f + h(T_s - T_f), \\
&(1-\phi)(\rho c)_s \frac{\partial T^s}{\partial t} = -(1-\phi)\frac{\partial Q_i}{\partial x_i} - h(T^s - T^f), \\
&\tau_s \frac{\partial Q_i}{\partial t} + Q_i = -k_s \frac{\partial T^s}{\partial x_i} \, ,
\end{aligned}
\tag{10.6}
$$

where τ_s is the solid thermal relaxation time. Equation $(10.6)_1$ is essentially Darcy's law, $(10.6)_2$ represents conservation of mass, $(10.6)_3$ is the energy balance equation for the fluid, $(10.6)_4$ is the energy balance equation for the solid, and $(10.6)_5$ is the Cattaneo equation for the solid heat flux. Equations (10.6) hold in the layer $\mathbb{R}^2 \times (0, d)$, and gravity acts in the downward direction. The boundary conditions are

$$v_3 = 0,\ z = 0, 1; \qquad T^f = T^s = T_L,\ z = 0; \qquad T^f = T^s = T_U,\ z = d, \tag{10.7}$$

where here T_L, T_U are constant temperatures with $T_L > T_U$.

The steady solution considered by Straughan (2013) is

$$\begin{aligned} &\bar{v} \equiv 0, \qquad \bar{T}^f = \bar{T}^s = -\beta z + T_L, \\ &\bar{Q}_1 = \bar{Q}_2 = 0, \quad \bar{Q}_3 = k_s\beta, \end{aligned} \tag{10.8}$$

where $\beta = (T_L - T_U)/d$ is the temperature gradient. Perturbations are introduced as

$$v_i = \bar{v}_i + u_i, \quad p = \bar{p} + \pi, \quad T^f = \bar{T}^f + \theta, \quad Q_i = \bar{Q}_i + q_i^s, \quad T^s = \bar{T}^s + \varphi.$$

The perturbation equations are non-dimensionalized by Straughan (2013) who introduces the non-dimensional parameters

$$Ra = R^2 = \frac{\rho_f^2 c_f^2 \beta d^2 K g \alpha}{\mu \phi k_f}, \qquad H = \frac{h d^2}{\phi k_f}, \qquad A = \frac{(\rho c)_s}{(\rho c)_f},$$
$$\gamma = \frac{\phi}{1-\phi}, \qquad \tau = \frac{\tau_s k_f}{\rho_f d^4},$$

where Ra and H are the Rayleigh number and the Rees number. The non-dimensional perturbation equations are

$$\begin{aligned} &u_i = -\frac{\partial \pi}{\partial x_i} + R\theta k_i, \\ &\frac{\partial u_i}{\partial x_i} = 0, \\ &\frac{\partial \theta}{\partial t} + u_i \frac{\partial \theta}{\partial x_i} = Rw + \Delta\theta + H(\varphi - \theta), \\ &A\frac{\partial \varphi}{\partial t} = -\frac{\partial q_i^s}{\partial x_i} - \gamma H(\varphi - \theta), \\ &\tau \frac{\partial q_i^s}{\partial t} + q_i^s = -\frac{\partial \varphi}{\partial x_i}. \end{aligned} \tag{10.9}$$

The boundary conditions are

$$w = 0, \quad \theta = 0, \quad \varphi = 0, \qquad z = 0, 1, \tag{10.10}$$

with the solution satisfying a plane tiling periodicity in the x, y directions.

Straughan (2013, 2015a,b) present extensive results for a solution to (10.9), (10.10), particularly when the solid skeleton is taken to be composed of copper oxide, or aluminium oxide, both of these materials being employed extensively in heat exchangers. The articles show that oscillatory convection is found in addition to oscillatory convection and they study in detail the transition from one to the other. One way in which (10.9) is different from the work in earlier chapters is that an energy stability analysis is possible even though we do not have a Mariano term present. The Laplacian of θ together with the H interaction terms provides a sufficient dissipation to make an energy stability analysis possible, see Straughan (2013).

10.2 Fluid inertia and anisotropy

Thermal convection in LTNE porous media with a Cattaneo effect in the solid has been studied recently by Hema et al. (2021), and by Capone and Gianfrani (2022). Both sets of writers employed a Brinkman theory for the porous medium, and both allowed for anisotropic permeability, and anisotropy in the fluid and solid heat fluxes. Additionally both sets of writers analysed stress free boundary conditions. Hema et al. (2021), also allowed for fluid inertia. Straughan (2016) has shown that Darcy theory is very different from Brinkman theory, and indeed, both may lead to different physical phenomena. We restrict attention to Darcy theory where we employ realistic boundary conditions and this theory is likely to be valid if the porosity is not overly high. We here extend the Darcy model given in equations (10.6) in two ways. The first is to include fluid inertia. The second is to allow the thermal conductivity of the solid to be horizontally isotropic. By that we mean the thermal conductivity will be the same in all directions in the x, y plane, but it will, in general, be different in the vertical direction. The inertia effect is seen to make the porous layer more stable. The horizontal isotropy in the solid skeleton is shown to lead to an interesting instability by oscillatory convection if the vertical solid heat flux is suitably larger than that in the horizontal directions. This is an effect which could be very useful in insulation theory or heat transfer theory since one could insert vertical strips in the porous material of a high thermal conductivity to enhance the vertical anisotropy.

The changes to equations (10.6) are that to include the inertia effect we incorporate a term of form $bv_{i,t}$ into $(10.6)_1$, for a constant b. To include horizontal isotropy of the solid thermal conductivity one modifies the thermal conductivity in equation $(10.6)_5$ to be a diagonal tensor. We do not include specific details of the relevant equations, but these will be shown below in the appropriate non-dimensional perturbation equations.

The governing equations for thermal convection in a Darcy porous material with LTNE effect and with the Cattaneo law pertaining in the solid are presented in section 10.1.1. A modification to allow for horizontal isotropy in the solid may be found in Capone and Gianfrani (2022), see also Straughan (2018). The configuration is such that the porous material occupies the horizontal layer between the planes $z = 0$ and $z = d$, with gravity acting downward. The upper plane is held at constant temperature, T_U, while the lower plane is maintained at constant temperature, T_L, with $T_L > T_U$. The basic steady state is as in (10.8), and consists of the temperature being linear in the vertical coordinate, z, with a constant vertical heat flux in the solid.

10.2.1 Vadasz effect

When fluid inertia is added to (10.9) the non-dimensional perturbation equations are

$$\begin{aligned} \frac{1}{Va} u_{i,t} + u_i &= -\pi_{,i} + R\theta k_i, \\ u_{i,i} &= 0, \\ \theta_{,t} + u_i \theta_{,i} &= Rw + \Delta\theta + H(\varphi - \theta), \\ A\varphi_{,t} &= -q_{i,i} + \gamma H(\theta - \varphi), \\ \tau q_{i,t} + q_i &= -\varphi_{,i}, \end{aligned} \tag{10.11}$$

these equations holding in the horizontal layer $(x, y) \in \mathbb{R}^2$, $z \in (0, 1)$, with $t > 0$. In (10.11) Va is the Vadasz number, see Straughan (2000).

10.2.2 Capone–Gianfrani effect

When a horizontally isotropic solid thermal conductivity is added to (10.9) the non-dimensional perturbation equations are

$$\begin{aligned} u_i &= -\pi_{,i} + R\theta k_i, \\ u_{i,i} &= 0, \\ \theta_{,t} + u_i \theta_{,i} &= Rw + \Delta\theta + H(\varphi - \theta), \\ A\varphi_{,t} &= -q_{i,i} + \gamma H(\theta - \varphi), \\ \tau q_{i,t} + q_i &= -\kappa^s_{ij}\varphi_{,j}, \end{aligned} \tag{10.12}$$

these equations holding in the horizontal layer $(x, y) \in \mathbb{R}^2$, $z \in (0, 1)$, with $t > 0$. In (10.12) κ^s_{ij} is the non-dimensional solid thermal conductivity. For horizontal isotropy one selects

$$\kappa^s_{ij} \equiv diag(\zeta, \zeta, 1), \tag{10.13}$$

as in Capone and Gianfrani (2022), where $\zeta = \kappa_h/\kappa_z$ where κ_z is the vertical thermal conductivity in the solid and κ_h is the solid thermal conductivity in any horizontal direction.

10.2.3 Stability

To analyse linear instability and nonlinear stability of a solution to either (10.11) or (10.12) it is convenient to put both effects into the system of equations and perform the mathematical analysis simultaneously. The numerical part of the work is then easily handled by appropriate choice of coefficients in the code.

The fully non-dimensional, nonlinear perturbation equations governing thermally convective disturbances with both Vadasz and Capone–Gianfrani terms may be shown to be

$$\begin{aligned}
&\frac{1}{Va}u_{i,t} + u_i = -\pi_{,i} + R\theta k_i,\\
&u_{i,i} = 0,\\
&\theta_{,t} + u_i\theta_{,i} = Rw + \Delta\theta + H(\varphi - \theta),\\
&A\varphi_{,t} = -q_{i,i} + \gamma H(\theta - \varphi),\\
&\tau q_{i,t} + q_i = -\kappa^s_{ij}\varphi_{,j}\,,
\end{aligned} \tag{10.14}$$

these equations holding in the horizontal layer $(x, y) \in \mathbb{R}^2$, $z \in (0, 1)$, with $t > 0$. To be precise, $Ra = R^2$ is the Rayleigh number,

$$Ra = \frac{g\alpha\beta d^2 K}{\nu\phi\kappa_f}\,.$$

The boundary conditions for a solution to (10.14) are

$$w = u_3 = 0, z = 0, 1; \theta = \varphi = 0, z = 0, 1; \tag{10.15}$$

and the solution satisfies a plane tiling periodicity in x and y. The cell forms which the solution may take are discussed at length in Chandrasekhar (1981, pages 43-52).

To develop a linear instability analysis we drop the nonlinear term from (10.14) and then write $u_i(\mathbf{x}, t) = e^{\sigma t}u_i(\mathbf{x})$, with a similar representation for θ, φ, q_i and π. We then take curl curl of the resulting equation from $(10.14)_1$, and introduce the function $h = q_{i,i}$. This leads one to solve the system

$$\begin{aligned}
&\Big(\frac{\sigma}{Va} + 1\Big)\Delta w = R\Delta^*\theta,\\
&\sigma\theta = \Delta\theta + Rw + H(\varphi - \theta),\\
&\sigma A\varphi + h + \gamma H(\varphi - \theta) = 0,\\
&\sigma\tau h + h + \zeta\Delta^*\varphi + \varphi_{zz} = 0,
\end{aligned} \tag{10.16}$$

where Δ^* is the horizontal Laplacian. One now employs a normal mode analysis, and after some calculation from (10.16) one may arrive at the fourth order equation in σ,

$$\begin{aligned}
&\Big(\frac{\sigma}{Va} + 1\Big)\Lambda\Big\{(\sigma\tau + 1)\Big[(\sigma + \Lambda)\sigma A + (\sigma + \Lambda)\gamma H + \sigma AH\Big]\Big\}\\
&+\Big(\frac{\sigma}{Va} + 1\Big)\Lambda\tilde{\Lambda}(\sigma + \Lambda + H)\\
&= R^2a^2\Big[\gamma H + \tilde{\Lambda} + \tau A\sigma^2 + \sigma(\tau\gamma H + A)\Big],
\end{aligned} \tag{10.17}$$

where $\Lambda = \pi^2 + a^2$, $\tilde{\Lambda} = \pi^2 + \zeta a^2$, with a being the wavenumber.

The stationary convection boundary is determined by taking $\sigma = 0$ in (10.17) and one sees that

$$R_{stat}^2 = \frac{\Lambda^2}{a^2} + \frac{\Lambda\tilde{\Lambda}H}{a^2(\gamma H + \tilde{\Lambda})}. \tag{10.18}$$

The critical Rayleigh and wavenumbers for stationary convection are determined by minimizing (10.18) in a^2.

To determine the oscillatory convection boundary we put $\sigma = i\omega$ in (10.17) and then take the real and imaginary parts of the result. This yields equations of the form

$$\begin{aligned} X_4\omega^4 - X_2\omega^2 + X_0 &= R^2a^2(-Y_2\omega^2 + Y_0),\\ -X_3\omega^2 + X_1 &= R^2a^2Y_1, \end{aligned} \tag{10.19}$$

where

$$\begin{aligned}
X_4 &= \frac{A\tau\Lambda}{Va},\\
X_3 &= \frac{\tau\Lambda}{Va}(A\Lambda + \gamma H + AH) + \Lambda A\Big(\tau + \frac{1}{Va}\Big),\\
X_2 &= \frac{\tau}{Va}\Lambda^2\gamma H + A\Lambda + \Lambda\Big(\tau + \frac{1}{Va}\Big)(A\Lambda + \gamma H + AH) + \frac{\Lambda\tilde{\Lambda}}{Va},\\
X_1 &= \Big(\tau + \frac{1}{Va}\Big)\Lambda^2\gamma H + \Lambda(A\Lambda + \gamma H + AH) + \Lambda\tilde{\Lambda}\Big(\frac{\Lambda + H}{Va}\Big) + \Lambda\tilde{\Lambda},\\
X_0 &= \Lambda^2\gamma H + \Lambda^2\tilde{\Lambda} + \Lambda\tilde{\Lambda}H,\\
Y_2 &= \tau A, \qquad Y_1 = \tau\gamma H + A, \qquad Y_0 = \gamma H + \tilde{\Lambda}.
\end{aligned}$$

From $(10.19)_2$ we find

$$\omega^2 = \frac{X_1 - R^2a^2Y_1}{X_3}, \tag{10.20}$$

and then from $(10.19)_1$ one may deduce

$$R_{osc}^2 = \frac{-\Gamma_2 \pm \sqrt{\Gamma_2^2 - 4\Gamma_0\Gamma_1}}{2\Gamma_1 a^2}, \tag{10.21}$$

where

$$\begin{aligned}
\Gamma_1 &= \frac{X_4Y_1^2}{X_3^2} - \frac{Y_1Y_2}{X_3},\\
\Gamma_2 &= -\frac{2X_1X_4Y_1}{X_3^2} + \frac{X_2Y_1}{X_3} - Y_0 + \frac{Y_2X_1}{X_3},\\
\Gamma_0 &= X_0 + \frac{X_1^2X_4}{X_3^2} - \frac{X_1X_2}{X_3}.
\end{aligned}$$

To find the critical Rayleigh number for oscillatory convection we minimize each solution of (10.21) checking carefully from (10.20) that $\omega_1^2 > 0$ and/or $\omega_2^2 > 0$, where the subscripts 1 and 2 refer to the + and - solutions in (10.21). We always found $\omega_2^2 < 0$, whereas $\omega_1^2 > 0$ for particular ranges of parameters. The numerical results reflect only solutions where $\omega_1^2 > 0$.

10.2.4 Numerical results for Vadasz effect

When $\tau = 0$ in (10.16) then oscillatory instabilities are impossible and only stationary convection is found, cf. Straughan (2006). Thus, the Vadasz effect alone is not responsible for oscillatory convection. When $Va = \infty$, i.e. fluid inertia is not present then oscillatory instability is found, see Straughan (2013). If $\tau = 0$ and one considers the classical theory of thermal convection in a porous material, where φ is not present, then Vadasz (1998) has shown that oscillatory convection is possible when the layer is rotating about a vertical axis, provided $0 < Va < \infty$. Straughan (2000) showed that when $Va = \infty$ then oscillatory convection does not occur when the layer is rotating. Hence, it seems reasonable to ask what effect does inertia have upon LTNE convection with a Cattaneo heat flux for the solid while the Vadasz inertia effect is also present.

Figure 10.1 shows the critical Rayleigh number behaviour as the Rees number H varies, for a selection of values of Va, with $\zeta = 1$. In each case oscillatory convection is found, but the most destabilizing is when $Va = \infty$. As Va decreases the oscillatory convection becomes more stable. Thus, increasing fluid inertia stabilizes the flow. We show stationary and oscillatory curves for Va having values of 10^{20}, 100, 50, 20, 15 and 10. The oscillatory curve decreases rapidly from the transition from stationary convection and tends to a horizontal limit in each case.

Hema et al. (2021) also found that Va decreasing stabilizes with Brinkman theory with two stress free surfaces. However, Straughan (2016) shows that Darcy theory is very different from Brinkman theory. Many porous materials used in the heat transfer industry are ceramics and are of small porosity where Darcy theory is likely to hold. Figure 10.1 also displays detail of Ra_{stat}, Ra_{osc}, for various values of Va, as H varies, with more detail than Hema et al. (2021).

10.2.5 Numerical results for the Capone–Gianfrani effect

We now concentrate on the horizontal isotropy effect in the heat flux for the solid. At first we consider $Va = \infty$, so there is no fluid inertia. This analysis is motivated by figure 5 of Capone and Gianfrani (2022). These writers additionally consider anisotropy in permeability and also of the heat flux in the fluid. However, the effect we believe is interesting is that displayed in their figure 5. We here consider a Darcy theory whereas Capone and Gianfrani (2022) have Brinkman theory with two stress free surfaces. We stress again that the Darcy theory is different from Brinkman theory in a non-trivial way as Straughan (2016) has demonstrated.

Figures 10.2 and 10.3 display the interesting behaviour we refer to. Figure 10.2 shows the variation of the critical Rayleigh number against ζ for values of $H = 10, 11, 12, 13$. We see that in each case for small ζ oscillatory convection dominates. As ζ is increased there is a transition where stationary convection takes over. Actually, if one continues along the oscillatory curve past the transition then increasing ζ leads to both $\omega_1^2 < 0$ and $\omega_2^2 < 0$. We deem this effect to be highly interesting as we believe the ζ values could be controlled in man made porous materials. Small values of ζ could be achieved for example by having vertical rods of metal with a high thermal conductivity such as copper. This would then allow one to control the type of convection witnessed and lower the instability threshold which could be highly useful in a heat transfer device where convective motion enhances heat transfer through the layer.

In figure 10.2 the parameter values are $A = 1, \tau = 0.5, \gamma = 1$. In figure 10.3 we choose $A = 1.42 \times 10^{-2}$ which is a value compatible with aluminium oxide, Al_2O_3, see Straughan (2013). We keep $\gamma = 1$ which means $\epsilon = 0.5$, a value typical of a heat transfer device. Figure 10.3 also displays critical Rayleigh numbers against ζ and the same behaviour is witnessed as in figure 10.2. However, the transition values are lower.

This is a very interesting situation where for ζ small enough oscillatory convection dominates and the transition to stationary convection occurs for higher ζ but also the values of Ra on the stationary convection curve are higher. There is a nonlinear relation between A, γ, τ and ζ which dictates where the transition lies, but the fact that oscillatory convection is found as ζ is lowered is important.

Figure 10.4 demonstrates the effect found in figures 10.2 and 10.3 as inertia is changed. The Rees number is held at $H = 9$ and we see that as Va decreases the oscillatory convection branch is increased meaning the layer is more stable. When we employed the parameters of figure 10.4 we found that when $Va = 10^2$ (or smaller) both ω_1^2 and ω_2^2 are negative. Thus, when fluid inertia and horizontally isotropic solid heat flux are both present it appears that decreasing Va stabilizes the layer.

We have so far only mentioned linear instability theory. One may perform a nonlinear energy stability analysis for equations (10.14). This was done for $Va = \infty, \zeta = 1$ by Straughan (2013), and by Capone and Gianfrani (2022) for $Va = \infty$, and Brinkman theory with two stress free surfaces. For (10.14) one utilizes an energy function of form

$$E(t) = \frac{1}{2Va}\|\mathbf{u}\|^2 + \frac{1}{2}\|\theta\|^2 + \frac{A}{2}\|\varphi\|^2 + \frac{\zeta}{2}(\|q_1\|^2 + \|q_2\|^2) + \frac{1}{2}\|q_3\|^2.$$

One finds in each situation that the global nonlinear stability threshold is $Ra = 4\pi^2$ as shown in figure 10.1 by the horizontal dotted line.

We have developed a stability analysis for a model of local thermal non-equilibrium convection in a saturated porous material where the heat flux of the solid skeleton obeys a Cattaneo equation rather than a Fourier one. This is likely to be the case in many situations where heat may travel rapidly. We concentrated on two effects and restricted analysis to Darcy theory. One problem was to analyse the effect fluid inertia will have on the critical Rayleigh number threshold which governs when convective motion may commence. The second problem was to investigate in detail how a horizontally isotropic heat flux in the solid affects the critical Rayleigh number threshold. We have shown that for both problems both stationary and oscillatory convection may occur. The most unstable situation for inertia is found to be when $Va = \infty$, i.e. when no fluid inertia is present. This is a case frequently chosen in porous media studies, i.e. where the fluid acceleration is neglected. It is also found that the horizonal isotropy of the solid heat flux is highly influential in dictating what type of convection occurs and what the convective motion thresholds are. When the vertical solid heat flux strongly dominates the horizontal one the convection found will be oscillatory.

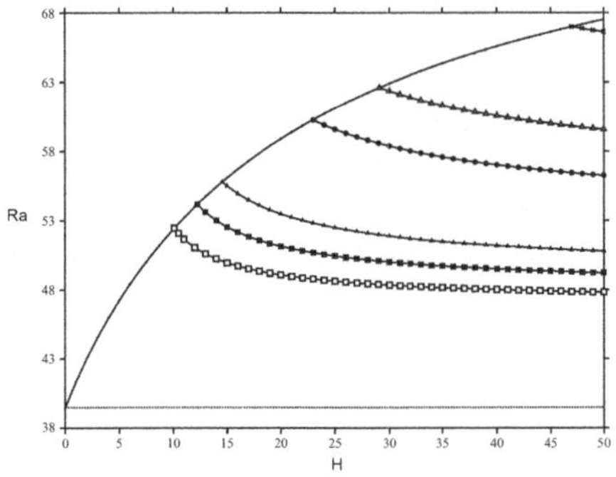

Fig. 10.1 Graph of critical Rayleigh number Ra versus H. Solid curve indicates stationary convection values. Dotted line at $Ra = 4\pi^2$ is the energy stability limit. The oscillatory curves correspond to □, $Va = 10^{20}$; ■, $Va = 100$; ▲, $Va = 50$; •, $Va = 20$; △, $Va = 15$; ×, $Va = 10$. $A = 1, \gamma = 1, \hat{\tau} = 0.5$. The transition values are $Va = 10^{20}, H = 10.105, Ra = 52.450$; $Va = 100, H = 12.249, Ra = 54.164$; $Va = 50, H = 14.555, Ra = 55.781$; $Va = 20, H = 23.00, Ra = 60.253$; $Va = 15, H = 29.078, Ra = 62.595$; $Va = 10, H = 47.00, Ra = 67.004$.

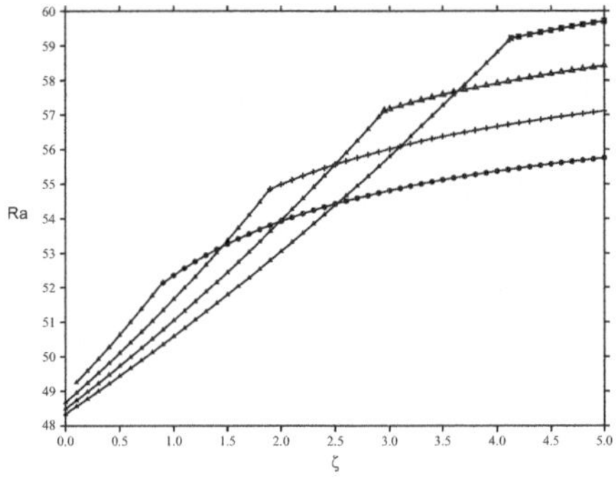

Fig. 10.2 Graph of critical Rayleigh number Ra versus ζ. Here $Va = 10^{20}, A = 1, \gamma = 1, \hat{\tau} = 0.5$. On each branch the curve starting in the lower left and designated with a ▲ represents oscillatory convection. The branches emanating to the right represent stationary convection. The stationary convection curves are •, $H = 10$; +, $H = 11$; △, $H = 12$; ■, $H = 13$. The transition values are $H = 10, \zeta = 0.95, Ra = 52.15$; $H = 11, \zeta = 1.90, Ra = 54.85$; $H = 12, \zeta = 2.95, Ra = 57.12$; $H = 13, \zeta = 4.13, Ra = 59.22$.

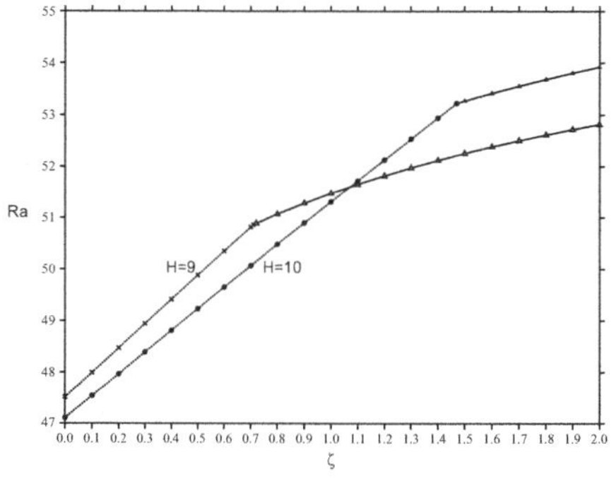

Fig. 10.3 Graph of critical Rayleigh number Ra versus ζ. $A = 1.42 \times 10^{-2}, \gamma = 1, \hat{\tau} = 0.5; Va = 10^{20}$. H values as on graphs. Transition values are $H = 9, \zeta = 0.715, Ra = 50.87$; $H = 10, \zeta = 1.47, Ra = 53.22$.

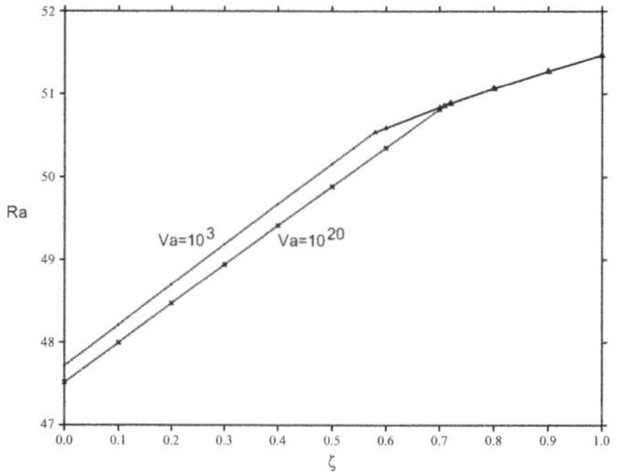

Fig. 10.4 Graph of critical Rayleigh number Ra versus ζ. $A = 1.42 \times 10^{-2}, \gamma = 1, \hat{\tau} = 0.5$; $H = 9$. Va values as on graphs. Transition values are $Va = 10^{20}, \zeta = 0.715, Ra = 50.87$; $Va = 10^3, \zeta = 0.58, Ra = 50.54$.

10.3 Mariano effect

In this section we return to the model of section 10.1.1 governed by equations (10.6), for Darcy LTNE convection with no fluid inertia, but with a Cattaneo effect in the solid skeleton. However, we here add a Mariano extra flux term to equation $(10.6)_4$. This means we add a term like $(1-\phi)\hat{\zeta}\Delta T^s$ to the right hand side of $(10.6)_4$. The boundary conditions and the steady solution are as (10.7) and (10.8). To analyse stability of the steady solution (10.8) we introduce perturbations and non-dimensionalize as in section 10.1.1.

The resulting non-dimensional perturbations equations have form

$$\begin{aligned}
u_i &= -\frac{\partial \pi}{\partial x_i} + R\theta k_i, \\
\frac{\partial u_i}{\partial x_i} &= 0, \\
\frac{\partial \theta}{\partial t} + u_i \frac{\partial \theta}{\partial x_i} &= Rw + \Delta\theta + H(\varphi - \theta), \\
A\frac{\partial \varphi}{\partial t} &= -\frac{\partial q_i^s}{\partial x_i} + \zeta\Delta\varphi - \gamma H(\varphi - \theta), \\
\tau\frac{\partial q_i^s}{\partial t} + q_i^s &= -\frac{\partial \varphi}{\partial x_i},
\end{aligned} \tag{10.22}$$

and the boundary conditions are (10.10). To analyse instability we linearize (10.22), assume a time dependence like $e^{\sigma t}$, and remove the pressure to arrive at the equations

$$\begin{aligned} &\Delta w - R\Delta^*\theta = 0, \\ &\sigma\theta = Rw + \Delta\theta + H(\varphi - \theta), \\ &A\sigma\varphi - \zeta\Delta\varphi + \Gamma + \gamma H(\varphi - \theta) = 0, \\ &(\tau\sigma + 1)\Gamma + \Delta\varphi = 0\,. \end{aligned} \tag{10.23}$$

After some calculation one may show that the instability problem which ensues from (10.23) reduces to solving the cubic equation

$$\begin{aligned} &\sigma^3\tau A\Lambda + \sigma^2\big[A\Lambda + \tau\Lambda(\zeta\Lambda + \gamma H) + \tau A(\Lambda^2 + H\Lambda)\big] \\ &\quad + \sigma\big\{\big[A + \tau(\zeta\Lambda + \gamma H)\big](\Lambda^2 + H\Lambda) + \zeta\Lambda^2 + \gamma H\Lambda + \Lambda^2\big\} \\ &\quad + (\zeta\Lambda + \gamma H + \Lambda)(\Lambda^2 + H\Lambda) - \sigma\tau H^2\gamma\Lambda - H^2\gamma\Lambda \\ &\quad = R^2a^2\big\{\tau A\sigma^2 + \sigma\big[A + \tau(\zeta\Lambda + \gamma H)\big] + \zeta\Lambda + \gamma H + \Lambda\big\}. \end{aligned} \tag{10.24}$$

The stationary convection boundary is found from (10.24) as

$$R^2_{stat} = \frac{(\zeta + 1)\Lambda^3 + \Lambda^2 H(\gamma + 1) + \zeta H\Lambda^2}{a^2(\Lambda + \zeta\Lambda + \gamma H)}\,. \tag{10.25}$$

The critical Rayleigh numbers and wavenumbers for stationary convection are found by minimizing (10.25) in a^2.

The oscillatory convection Rayleigh numbers are found by putting $\sigma = i\omega$ in (10.24) and then taking the real and imaginary parts of the resulting equation. This results in two equations of form

$$\begin{aligned} &-\omega^2 E_1 + E_2 = -\omega^2\tau A R^2 a^2 + E_3 R^2 a^2, \\ &-\omega^2 F_1 + F_2 = R^2 a^2 F_3\,, \end{aligned} \tag{10.26}$$

where

$$\begin{aligned} E_1 &= A\Lambda + \tau\Lambda(\zeta\Lambda + \gamma H) + \tau A(\Lambda^2 + H\Lambda), \\ E_2 &= (\zeta + 1)(\Lambda^3 + H\Lambda^2) + \gamma H\Lambda^2, \\ E_3 &= \zeta\Lambda + \gamma H + \Lambda, \\ F_1 &= \tau A\Lambda, \\ F_2 &= A\Lambda(\Lambda + H) + \tau\Lambda^2(\zeta\Lambda + \gamma H) + (\zeta + 1)\Lambda^2 + \gamma H\Lambda + \tau\zeta H\Lambda, \\ F_3 &= A + \tau(\zeta\Lambda + \gamma H). \end{aligned}$$

It is seen that

$$\omega^2 = \frac{F_2 - R^2 a^2 F_3}{F_1}\,. \tag{10.27}$$

Equations (10.26) lead to the quadratic equation

$$A_1(R^2a^2)^2 - A_2R^2a^2 + A_3 = 0, \tag{10.28}$$

where

$$A_1 = \frac{\tau AF_3}{F_1}, \qquad A_2 = \frac{E_1F_3}{F_1} + \frac{\tau AF_2}{F_1} - E_3, \qquad A_3 = \frac{E_1F_2}{F_1} - E_2.$$

The oscillatory convection boundary is then found from

$$R^2 = \frac{A_2 \pm \sqrt{A_2^2 - 4A_1A_3}}{2a^2A_1}. \tag{10.29}$$

To find the critical oscillatory convection Rayleigh and wave numbers one must *carefully* minimize (10.29) in a^2, checking that $\omega^2 > 0$ for the roots using (10.27).

Numerical results are provided in table 10.1 and figure 10.5, where $A = 1, \gamma = 1, \tau = 0.5$. For the values we have investigated stationary convection is witnessed for H small enough, for all ζ analysed. Once H is large enough there is a transition to oscillatory convection as shown in curves a - d in figure 10.5. The behaviour of the oscillatory convection curve is interesting although somewhat involved. For values of ζ small enough, here for $\zeta = 0.1, 0.2, \ldots, 1$, the oscillatory convection curve descends from the transition point and reaches a minimum, and thereafter increases. For example, on curve c the minimum is reached when $H = 11$. For values of ζ higher than 1, the oscillatory convection curve always increases when increasing H after the transition point. It thus appears that when ζ is small the Mariano effect is small and after the oscillatory convection mechanism which is induced entirely by the Cattaneo relaxation effect in the solid acts to destabilize the steady solution. Once the Mariano effect is sufficiently large it appears to dominate the behaviour, here for $\zeta = 2, 3, \ldots, 10$, the last curve being labelled d in figure 10.5. For a fixed value of H we see that the Mariano effect destabilizes the steady solution when H is sufficiently small, as may be seen from figure 10.5 at the value $H = 11$. The curves shown are for $\zeta = 0, 0.3, 1$ and 10, and as ζ increases the oscillatory convection instability threshold decreases. Since the Mariano term, the ζ term, is usually stabilizing as it is a diffusion term in classical porous media theory, the Cattaneo effect is inducing a counter intuitive behaviour.

The behaviour shown in figure 10.5 holds for all of the values of $\zeta = 0, 0.1, \ldots, 10$, which are given in table 10.1 where the transition values from stationary to oscillatory convection are displayed.

10.4 Bidisperse convection

In this section we present a model for thermal convection in what is called a bidisperse, or double porosity, porous medium. In such a material there are the usual pores which are called macro pores, but there are also cracks or fissures in the solid skeleton which give rise to pores on a much smaller scale, the micro pores. The

Table 10.1 Transition values, $Ra_{stat} = Ra_{osc}$, changing from stationary to oscillatory convection. $A = 1, \gamma = 1, \tau = 0.5$.

Ra	a^2_{stat}	a^2_{osc}	H	ζ
52.4504	11.81	10.96	10.106	0
51.6278	11.85	11.29	8.832	0.1
50.8052	11.84	11.54	7.793	0.2
50.0410	11.81	11.68	6.958	0.3
49.3614	11.76	11.75	6.290	0.4
48.7736	11.70	11.76	5.756	0.5
48.2720	11.65	11.75	5.327	0.6
47.8427	11.60	11.72	4.977	0.7
47.4758	11.55	11.68	4.689	0.8
47.1604	11.51	11.64	4.449	0.9
46.8896	11.47	11.61	4.248	1
45.4255	11.23	11.35	3.2348	2
44.8488	11.12	11.22	2.8652	3
44.5448	11.07	11.15	2.6763	4
44.3578	11.03	11.11	2.5620	5
44.2314	11.00	11.08	2.4855	6
44.1402	10.99	11.06	2.4308	7
44.0714	10.97	11.04	2.3897	8
44.0176	10.96	11.03	2.3577	9
43.9744	10.95	11.02	2.3321	10

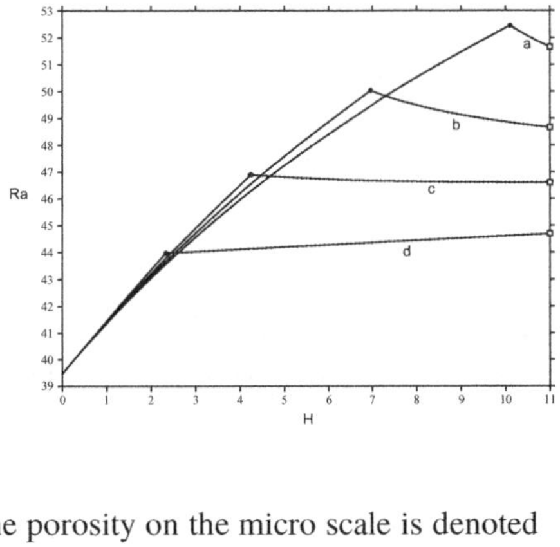

Fig. 10.5 Graph of critical Rayleigh number Ra versus H. $A = 1, \gamma = 1, \hat{\tau} = 0.5$. Curve a is for $\zeta = 0$, curve b is for $\zeta = 0.3$, curve c is for $\zeta = 1$, curve d is for $\zeta = 10$. Transition values are for $\zeta = 0$, $H = 10.106$, $Ra = 52.4505$, for $\zeta = 0.3$, $H = 6.958$, $Ra = 50.0410$, for $\zeta = 1$, $H = 4.248$, $Ra = 46.8896$, for $\zeta = 10$, $H = 2.332$, $Ra = 43.9744$. The Rayleigh number values when $H = 11$ are $Ra = 51.6612$, for $\zeta = 0$, $Ra = 48.6673$, for $\zeta = 0.3$, $Ra = 46.6007$, for $\zeta = 1$, $Ra = 44.6897$, for $\zeta = 10$.

macro porosity is denoted by ϕ whereas the porosity on the micro scale is denoted by ϵ.

We believe the first model for thermal convection in a bidisperse porous material is due to Nield and Kuznetsov (2006), see also Nield and Bejan (2013), or Straughan (2015a, chapter 13). Thermal convection in bidispersive porous media is now a very active area of research driven by many practical applications, see e.g. Gentile and Straughan (2017, 2020), Straughan (2019), Capone et al. (2020b,a), Capone et al. (2021), Capone et al. (2022), Capone et al. (2023), Saravanan and Vigneshwaran (2020), Siddabasappa et al. (2023), Eltayeb (2023), Tripathi et al. (2024), Jia and Jian (2024), Franchi et al. (2024), and the many references therein.

The original model of Nield and Kuznetsov (2006) had separate velocities, temperatures, and pressures in the macro pores and in the micro pores, and they employed Brinkman theory in both macro and micro pores. LTNE effects have been incorporated into bidisperse porous media, and our goal here is to present a model for thermal convection with LTNE effects in a bidisperse porous medium allowing for a Cattaneo heat flow relaxation effect in the solid skeleton.

To achieve our goal we begin with the model of Franchi et al. (2017) who used Darcy theory, but had separate velocities in the macro and micro pores, and they had independent temperatures in the macro and micro pores, and in the solid skeleton. We now generalize their model as follows. Let T^f, T^p, T^s be the temperatures in the macro pores, the micro pores, and in the solid, and let V_i^f and V_i^p be the actual fluid velocities in the macro pores and in the micro pores. Let p^f and p^p be the pressures in the macro pores and in the micro pores, and suppose Q_i^s is the heat flux in the solid skeleton of the porous material. Using Franchi et al. (2017) as a starting point we write the governing equations as the equations of momentum in the macro and micro pores,

$$
\begin{aligned}
&-\frac{\mu}{K_f}U_i^f - \zeta(U_i^f - U_i^p) - p_{,i}^f - \frac{g_i\rho_0\alpha\phi}{D}T^f - \frac{g_i\rho_0\alpha(1-\phi)\epsilon}{D}T^p = 0,\\
&U_{i,i}^f = 0,\\
&-\frac{\mu}{K_p}U_i^f - \zeta(U_i^p - U_i^f) - p_{,i}^p - \frac{g_i\rho_0\alpha\phi}{D}T^f - \frac{g_i\rho_0\alpha(1-\phi)\epsilon}{D}T^p = 0,\\
&U_{i,i}^p = 0,
\end{aligned}
\tag{10.30}
$$

where we employ the notation here and below,

$$
\begin{aligned}
&\epsilon_1 = (1-\epsilon)(1-\phi), \qquad \epsilon_2 = (1-\phi)\epsilon, \qquad D = \phi + (1-\phi)\epsilon,\\
&g_i = -k_i g, \qquad U_i^f = \phi V_i^f, \qquad U_i^p = \epsilon(1-\phi)V_i^p,
\end{aligned}
$$

g being gravity, $\mathbf{k} = (0,0,1)$, and f, p, s refer to the macro pores, micro pores, and solid, respectively. In addition, α is the thermal expansion coefficient of the fluid, ρ_0 is a constant reference density, μ is the dynamic viscosity of the fluid. The velocities U_i^f, U_i^p are the pore averaged velocities in the continuum model of the porous medium. There are the equations of energy balance in the macro and micro pores, namely,

$$
\begin{aligned}
&\phi(\rho c)_f T_{,t}^f + \phi(\rho c)_f V_i^f T_{,i}^f = \phi\kappa_f \Delta T^f + h(T^p - T^f) + s_1(T^s - T^f),\\
&\epsilon_2(\rho c)_p T_{,t}^p + \epsilon_2(\rho c)_p V_i^p T_{,i}^p = \epsilon_2\kappa_p \Delta T^p + h(T^f - T^p) + s_2(T^s - T^p),
\end{aligned}
\tag{10.31}
$$

where h, s_1, s_2 are interaction coefficients due to LTNE effects, ρ, c are specific heats referred to the fluid in the macro or micro pores, or in the solid depending on the subscript or superscript f, p or s. The coefficients κ are thermal conductivities again depending on the sub or superscript. The balance of energy equation for the solid skeleton is

$$\epsilon_1(\rho c)_s T^s_{,t} = -\epsilon_1 Q^s_{i,i} + s_1(T^f - T^s) + s_2(T^p - T^s), \tag{10.32}$$

and the equation of Cattaneo for the solid is

$$\hat{\tau} Q^s_{i,t} + Q^s_i = -\kappa_s T^s_{,i}, \tag{10.33}$$

where $\hat{\tau}$ is the relaxation coefficient.

Equations (10.30), (10.31), (10.32), (10.33), represent a general model for heat flow but we here make a further assumption that $T^f = T^p = T$. This means that the temperatures of the fluid in the macro and micro pores is the same, and we expect this to be true in most situations, cf. Franchi et al. (2024). With this simplification equations (10.30) - (10.33) reduce to the following groups of equations. The momentum and continuity of mass equations

$$\begin{aligned} &-\frac{\mu}{K_f}U^f_i - \zeta(U^f_i - U^p_i) - p^f_{,i} - g_i\rho_0\alpha T = 0, \\ &U^f_{i,i} = 0, \\ &-\frac{\mu}{K_p}U^f_i - \zeta(U^p_i - U^f_i) - p^p_{,i} - g_i\rho_0\alpha T = 0, \\ &U^p_{i,i} = 0. \end{aligned} \tag{10.34}$$

The equation for the temperature of the fluid

$$D(\rho c)_f T_{,t} + (U^f_i + U^p_i)(\rho c)_f T_{,i} = D\kappa_f \Delta T + (s_1 + s_2)(T^s - T). \tag{10.35}$$

The equation for the temperature of the solid

$$\epsilon_1(\rho c)_s T^s_{,t} = -\epsilon_1 Q^s_{i,i} + (s_1 + s_2)(T - T^s). \tag{10.36}$$

In addition one has the Cattaneo equation (10.33). Thus, the system of equations for our model of LTNE bidisperse convection with a Cattaneo effect in the solid is (10.34), (10.35), (10.36) and (10.33) which we shall refer to as equations C. If we set $\hat{\tau} = 0$, i.e. if we neglect the Cattaneo effect, this system is analysed in some detail in Franchi et al. (2024). In fact they find that for thermal convection the perturbation system is represented by a symmetric linear operator and they are able to deduce equivalence between the linear instability and nonlinear stability boundaries. This is *not* true here. The perturbation equations which follow from equations C do not represent a symmetric operator and oscillatory instability is possible.

For thermal convection equations C hold in the spatial domain $\{(x, y) \in \mathbb{R}^2\} \times \{z \in (0, d)\}$ with gravity acting in the downward (negative z) direction. The boundary conditions are no flow out of the boundaries and

$$T = T^s = T_U, \quad z = d; \qquad T = T^s = T_L, \quad z = 0;$$

with $T_L > T_U$, both constants. The leads to the steady solution

$$\bar{T} = \bar{T}^s = -\beta z + T_L, \qquad \bar{v}_i = 0, \qquad \bar{\mathbf{q}} = (0, 0, \kappa_s \beta), \tag{10.37}$$

where $\beta = (T_L - T_U)/d$.

To analyse instability of solution (10.37) we introduce perturbations $u_i^f, u_i^s, \theta, \theta^s$, q_i^s, π^f, π^p and non-dimensionalize these equations with length scale d, velocity scale U, time scale $\mathcal{T}$, pressure scale P, temperature scale $T^\sharp$, where

$$U = \frac{D\kappa_f}{d(\rho c)_f}, \qquad T^\sharp = U\sqrt{\frac{\mu\beta(\rho c)_f d^2}{D\kappa_f \alpha g \rho_0 K_f}},$$
$$P = \frac{\mu U d}{K_f}, \qquad \mathcal{T} = \frac{(\rho c)_f d^2}{\kappa_f}.$$

The solid heat flux q_i^s is scaled with $Q^\sharp = \kappa_s T^\sharp / d$ and we introduce the Rayleigh number $Ra = R^2$, the Rees number S, the relative permeability K_r, and the coefficients a, b, τ, ξ and γ by

$$Ra = R^2 = \frac{\alpha g d^2 \beta K_f}{(\mu/\rho_0)[\kappa_f/(\rho c)_f]D},$$
$$S = \frac{(s_1 + s_2)d^2}{D\kappa_f}, \qquad K_r = \frac{K_f}{K_p}, \qquad a = \frac{\epsilon_1(\rho c)_s}{\mathcal{T}},$$
$$b = \frac{\epsilon_1 \kappa_s}{d^2}, \qquad \gamma = \frac{D\kappa_f}{d^2}, \qquad \tau = \frac{\hat{\tau}}{\mathcal{T}}, \qquad \xi = \frac{\zeta K_f}{\mu}.$$

The non-dimensional perturbation equations may then be written as

$$\begin{aligned}
&- u_i^f - \xi(u_i^f - u_i^p) - \pi^f_{,i} + R\theta k_i = 0, \qquad u^f_{i,i} = 0,\\
&- K_r u_i^p - \xi(u_i^p - u_i^f) - \pi^f_{,i} + R\theta k_i = 0, \qquad u^p_{i,i} = 0,\\
&\theta_{,t} + (u_i^f + u_i^p)\theta_{,i} = R(w^f + w^p) + \Delta\theta + S(\theta^s - \theta),\\
&a\theta^s_{,t} = -b q^s_{i,i} + \gamma S(\theta - \theta^s),\\
&\tau q^s_{i,t} + q_i^s = -\theta^s_{,i},
\end{aligned} \tag{10.38}$$

where $w^f = u_3^f, w^p = u_3^p$, and (10.38) hold on the domain $\mathbb{R}^2 \times \{z \in (0,1)\}$, for $t > 0$. The boundary conditions are that $u_i^f, u_i^p, \theta, \theta^s, q_i^s, \pi^f, \pi^p$ are periodic in x, y, and on $z = 0, 1$ they satisfy

$$w^f = w^p = 0, \qquad \theta = \theta^s = 0, \qquad z = 0, 1. \tag{10.39}$$

To develop an instability analysis one removes π^f and π^p and linearizes, then one assumes a time dependence like $e^{\sigma t}$. The system of equations which subsequently results is

$$
\begin{aligned}
&\Delta w^f + \xi(\Delta w^f - \Delta w^p) - R\Delta^*\theta = 0,\\
&K_r\Delta w^p + \xi(\Delta w^p - \Delta w^f) - R\Delta^*\theta = 0,\\
&\sigma\theta = R(w^f + w^p) + \Delta\theta + S(\theta^s - \theta),\\
&\sigma a\theta^s = -b\Gamma + \gamma S(\theta - \theta^s),\\
&\sigma\tau\Gamma = -\Gamma - \Delta\theta^s,
\end{aligned}
\tag{10.40}
$$

where $\Gamma = q^s_{i,i}$. Due to the boundary conditions we may develop a normal mode solution and we proceed to see that the growth rate σ satisfies the cubic equation

$$
\begin{aligned}
&\left[\tau a\sigma^2 + \sigma(a + \gamma\tau S) + b\Lambda + \gamma S\right]\\
&\quad\times\left\{(\Lambda + S + \sigma)\Lambda^2\left[\xi^2 + (1 + K_r)\xi\right] - R^2a^2\Lambda(1 + 4\xi + K_r)\right\}\\
&\quad - \gamma S^2\Lambda^2(\tau\sigma + 1)\left[\xi^2 + (1 + K_r)\xi\right] = 0.
\end{aligned}
\tag{10.41}
$$

The stationary convection boundary is found from (10.41) by putting $\sigma = 0$, and this boundary is the same as the analogous one found in Franchi et al. (2024), and this is

$$
R^2 = \frac{\Lambda^2}{a^2}\left(\frac{b(\Lambda + S) + S}{b\Lambda + S}\right)\left(\frac{K_r + K_r\xi + \xi}{1 + 4\xi + K_r}\right). \tag{10.42}
$$

Franchi et al. (2024) discuss in detail numerical results from (10.42) dealing with practical cases such as water for the fluid and glass beads for the solid, or a heat exchanger oil for the fluid with the solid being a high thermal conductivity copper. We believe it will be interesting to analyse (10.40) to include oscillatory convection which arises from the model presented here.

References

Al Ghoul, M. (2004). Generalized hydrodynamics of reaction - diffusion systems and dissipative structures. *Phil. Trans. Roy. Soc. London A*, 362:1567–1581.

Alfifi, H. (2022). Stability analysis for Schnakenberg reaction - diffusion model with gene expression time delay. *Chaos, Solitons and Fractals*, 155:111730.

Allocca, V., Colantuono, P., Collela, A., Piacentini, S. M., and Piscopo, V. (2022). Hydraulic properties of ignimbrites: matrix and fracture properties in two pyroclastic flow deposits from Cimino - Vico volcanoes (Italy). *Bull. Engng. Geology and the Environment*, 81:221.

Andres, M. and Pinnau, R. (2019). The Cattaneo model in the context of thermoablation of liver tumors. *Proc. Appl. Math. Mech.*, 19:e201900241.

Andres, M. and Pinnau, R. (2022). The Cattaneo model for laser - induced thermotherapy; identification of the blood - perfusion rate. In Pinnau, R., Gauger, N. R., and Klar, A., editors, *Modeling, simulation and optimization in the health - and energy - sector*, volume 14, pages 25–41. Springer, Cham. SEMA SIMAI Springer Series.

Angeles, F. (2023). Hyperbolic systems of quasilinear equations in compressible fluid dynamics with an objective Cattaneo - type extension for the heat flux. *Mech. Research Comm.*, 130:104103.

Ashwin, T., Narasimham, G. S. V. L., and Jacob, S. (2010). CFD analysis of high frequency miniature pulse tube refrigerators for space applications with thermal non - equilibrium model. *Applied Thermal Engineering*, 30:152–166.

Baek, J. Y., Park, B. H., Rau, G. C., and Lee, K. K. (2022). Experimental evidence for local thermal non - equilibrium during heat transport in sand representative of natural conditions. *J. Hydrology*, 608:127589.

Bagdassarov, N. S. and Fradkov, A. S. (1993). Evolution of double diffusive convection in a felsic magma chamber. *J. Vulcanology and Geothermal Research*, 54:291–308.

Banu, N. and Rees, D. A. S. (2002). Onset of Darcy - Bénard convection using a thermal non-equilibrium model. *Int. J. Heat Mass Transfer*, 45:2221–2228.

Barbera, E., Curro, C., and Valenti, G. (2008). A hyperbolic reaction - diffusion model for the hantavirus infection. *Math. Meth. Appl. Sci.*, 31:481–499.

Barbera, E., Currò, C., and Valenti, G. (2010). A hyperbolic model for the effects of urbanization on air pollution. *Appl. Math. Modelling*, 34:2192–2202.

Barbera, E. and Pollino, A. (2023). A hyperbolic reaction - diffusion model of chronic wasting disease. *Ricerche di Matematica*, pages 1–13. https://doi.org/10.1007/s11587-023-00831-8.

Barbera, E. and Valenti, G. (2018). Wave features of a hyperbolic reaction - diffusion model for chemotaxis. *Wave Motion*, 78:116–131.

Bargmann, S., Greve, R., and Steinmann, P. (2008). Simulation of cryovolcanism on Saturn's moon Enceladus with the Green - Naghdi theory of thermoelasticity. *Bulletin of Glaciological Research*, 26:23–32.

Barletta, A. (2019). *Routes to absolute instability in porous media*. Springer, New York.

Barletta, A. (2021). Spatially developing modes: The Darcy - Bénard problem revisited. *Physics*, 3:549–562.

Barletta, A. (2022). The Boussinesq approximation for buoyant flows. *Mech. Research Comm.*, 124:103939.

Barletta, A., Celli, M., and Rees, D. A. S. (2023). On the use and misuse of the Oberbeck - Boussinesq approximation. *Physics*, 5:298–309.

Batchelor, G. (1967). *An introduction to fluid dynamics*. Cambridge University Press, Cambridge.

Bénard, H. (1900). Les tourbillons cellulaires dans une nappe liquide. *Revue Gén. Sci. Pure Appl.*, 11:1261–1271.

Berezovski, A. (2016). On the influence of microstructure on heat conduction in solids. *Int. J. Heat Mass Transfer*, 103:516–520.

Berezovski, A. (2019). Internal variables representation of generalized heat equations. *Continuum Mech. Thermodyn.*, 31:1733–1741.

Berselli, L. C. and Bisconti, L. (2012). On ther structural stability of the Euler - Voigt and Navier - Stokes - Voigt models. *Nonlinear Analysis*, 75:117–130.

Bidin, B. and Rees, D. A. S. (2020). Pattern selection for Darcy - Bénard convection with local thermal nonequilibrium. *Int. J. Heat Mass Transfer*, 153:119539.

Bissell, J. J. (2015). On oscillatory convection with the Cattaneo - Christov hyperbolic heat flow model. *Proc. Roy. Soc. London A*, 471:20140845.

Bissell, J. J. (2016). Thermal convection in a magnetized conducting fluid with the Cattaneo - Christov heat flow model. *Proc. Roy. Soc. London A*, 472:20160649.

Blauth, S., Andres, M., Pinnau, R., and Totzcek, C. (2019). Optimal control and asymptotic analysis of the Cattaneo equation. *Proc. Appl. Math. Mech.*, 19:e201900184.

Boltzmann, L. (1874). Zur Theorie der Elastischen Nachwirkung. *Sitzungsber. Kaiserlich Akad. Wiss. (Wien) Math. Naturwiss. Classe*, 70:275–300.

Böttcher, F., Casasso, A., Gótzl, G., and Zosseder, K. (2019). TAP - Thermal aquifer Potential; A quantitative method to assess the spatial potential for the thermal use of ground water. *Renewable Energy*, 142:85–95.

Boussinesq, M. J. (1897). *Théorie de l'ecoulement tourbillonnant et tumultueux des liquides dans les lits rectilignes a grande section*. Gauthier - Villars, Paris.

Breugem, W. P. and Rees, D. A. S. (2006). A derivation of the volume-averaged Boussinesq equations for flow in porous media with viscous dissipation. *Trans. Porous Media*, 63:1–12.

Capone, F., De Luca, R., Fiorentino, L., and Massa, G. (2023). Bidisperse convection under the action of an internal heat source. *Int. J. Nonlinear Mech.*, 150:104360.

Capone, F., De Luca, R., and Gentile, M. (2020a). Coriolis effect on thermal convection in a rotating bidispersive porous layer. *Proc. Roy. Soc. London A*, 476:20190875.

Capone, F., De Luca, R., and Gentile, M. (2020b). Penetrative convection in rotating anisotropic bidispersive porous layers. *Mech. Res. Comm.*, 110:103601.

Capone, F., De Luca, R., and Massa, G. (2022). The onset of double diffusive convection in a rotating bidisperse porous medium. *Eur. Phys. J. Plus*, 137:1034.

Capone, F., De Luca, R., and Vitiello, M. (2019). Double - diffusive Soret convection phenomenon in porous medium: effect of Vadasz inertia term. *Ricerche di Matematica*, 68:581–595.

Capone, F. and Gentile, M. (2018). Sharp stability results in LTNE rotating anisotropic porous layer. *Int. J. Heat Thermal Science*, 134:661–664.

Capone, F. and Gianfrani, J. A. (2022). Onset of convection in LTNE Darcy - Bénard anisotropic layer: Cattaneo effect in the solid. *Int. J. Nonlinear Mech.*, 139:103889.

Capone, F., Massa, G., and Gentile, M. (2021). The onset of thermal convection in anisotropic and rotating bidisperse porous media. *ZAMP*, 72:169.

Capriz, G. and Mariano, P. M. (2014). Objective fluxes in a multi-scale continuum description of sparse medium dynamics. *Physica A*, 415:354–365.

Capriz, G., Wilmanski, K., and Mariano, P. M. (2021). Exact and appropriate Maxwell - Cattaneo type descriptions of heat conduction: a comparative analysis. *Int. J. Heat Mass Transfer*, 175:121362.

Carillo, S. and Jordan, P. M. (2020). On the propagation of temperature - rate waves and travelling waves in rigid conductors of the Graffi - Franchi - Straughan type. *Math. Comp. Simulation*, 176:120–133.

Carlomagno, I., Di Domenico, M., and Sellitto, A. (2021). High order fluxes in heat transfer with phonons and electrons: application to wave propagation. *Proc. Roy. Soc. London A*, 477:20210392.

Cattaneo, C. (1948). Sulla conduzione del calore. *Atti Sem. Mat. Fis. Modena*, 3:83–101.

Chandrasekhar, S. (1981). *Hydrodynamic and hydromagnetic stability*. Dover, New York.

Chirita, S. (2018). High - order effects of thermal lagging in deformable conductors. *Int. J. Heat Mass Transfer*, 127:965–974.

Christov, C. I. (2007). On a higher gradient generalization of Fourier's law of heat conduction. In *Amer. Inst. Phys. Conf. Proc.*, volume 946, pages 11–22.

Christov, C. I. (2009). On frame indifferent formulation of the Maxwell - Cattaneo model of finite - speed heat conduction. *Mech. Res. Comm.*, 36:481–486.

Christov, C. I. and Jordan, P. M. (2005). Heat conduction paradox involving second - sound propagation in moving media. *Phys. Rev. Letters*, 94:154301.

Christov, I. C. (2016). Nonlinear acoustics and shock formation in lossless barotropic Green - Naghdi fluids. *Evolution Equations and Control Theory*, 5:349–365.

Christov, I. C. (2019). On a C-integrable equation for second sound propagation in heated dielectrics. *Evolution Equations and Control Theory*, 8:57–72.

Cimatti, G. (2024). A theorem of existence of the solution for the nonlocal formulation of the problem of microwave heating. *Annali dell'Università di Ferrara*, 70:533–545.

Cimmelli, V. A. (2021). Local versus nonlocal continuum theories of nonequilibrium thermodynamics: the Guyer - Krumhansl equation as an example. *ZAMP*, 72:195.

Cimmelli, V. A., Sellitto, A., and Jou, D. (2010). Nonequilibrium temperatures, heat waves, and nonlinear heat transport equations. *Phys. Rev. B*, 81:054301.

Coleman, B. D. and Mizel, V. J. (1963). Thermodynamics and departures from fourier's law of heat conduction. *Arch. Rational Mech. Anal.*, 13:245–261.

Consolo, G., Currò, C., Grifó, G., and Valenti, G. (2022). Oscillatory periodic pattern dynamics in hyperbolic reaction - advection - diffusion models. *Phys. Rev. E*, 105:034206.

Consolo, G., Currò, C., and Valenti, G. (2017). Pattern formation and modulation in a hyperbolic vegetation model for semi - arid environments. *Appl. Math. Modelling*, 43:372–392.

Consolo, G., Currò, C., and Valenti, G. (2019). Supercritical and subcritical Turing pattern formation in a hyperbolic vegetation model for flat arid environments. *Physica D*, 398:141–163.

Consolo, G., Currò, C., and Valenti, G. (2020). Turing vegetation patterns in a generalized hyperbolic Klausmeier model. *Math. Meth. Appl. Sci.*, 43:10474–10489.

Currò, C. and Valenti, G. (2021). Pattern formation in hyperbolic models with cross - diffusion: theory and applications. *Physica D*, 418:132846.

Dai, W. Z., Wang, H. J., Jordan, P. M., Mickens, R. E., and Bejan, A. (2008). A mathematical model for skin burn injury induced by radiation heating. *Int. J. Heat Mass Transfer*, 51:5497–5510.

Damázio, P. D., Manholi, P., and Silvestre, A. L. (2016). L^q - theory of the Kelvin - Voigt equations in bounded domains. *J. Differential Equations*, 260:8242–8260.

Damm, D. L. and Fedorov, A. G. (2006). Local thermal non - equilibrium effects in porous electrodes of the hydrogen - fueled SOFC. *J. Power Sources*, 159:1153–1157.

De Campos, C. P., Dingwell, D. B., and Fehr, K. T. (2005). Evidence of double diffusive convection in the evolution of the Phlegrei Fields reservoir. *Geophys. Res. Abstracts*, 7:01132.

Deepika, N. (2018). Linear and nonlinear stability of double diffusive convection with the Soret effect. *Trans. Porous Media*, 121:93–108.

Degiovanni, M., Marzocchi, A., and Mastaglio, S. (2021). Regularity for the second grade Navier - Stokes equations in exterior domains. In Bodnar, T., Galdi, G. P., and Necasova, S., editors, *Waves in Flows*, pages 181–202. Birkhauser, Cham. Lecture Notes in Mathematical Fluid Mechanics.

Dell'Oro, F. and Pata, V. (2023). A hierarchy of heat conduction laws. *Discrete and Continuous Dynamical Systems S*, 16:2636–2648.

Dineshkumar, P. and Raja, M. (2022). An experimental study on trapezoidal salt gradient solar pond using magnesium sulfate ($MgSO_4$) salt and coal cinder. *J. Thermal Anal. Calorimetry*, 147:10525–10532.

Dolak, Y. and Hillen, T. (2003). Cattaneo models for chemosensitive movement. *J. Math. Biology*, 46:461–478.

Dongarra, J. J., Straughan, B., and Walker, D. W. (1996). Chebyshev tau - QZ algorithm methods for calculating spectra of hydrodynamic stability problems. *Appl. Numer. Math.*, 22:399–435.

Elshehabey, H. M., Aly, A. M., Lee, S. W., and Colak, A. (2024). Integrating artificial intelligence with numerical simulations of Cattaneo - Christov heat flux on thermosolutal convection of nano - enhanced phase change materials in Bézier - annulus. *J. Energy Storage*, 82:110496.

Eltayeb, I. A. (2015). Stability of porous Bénard - Brinkman layer in local thermal non - equilibrium with Cattaneo effects in the solid. *Int. J. Thermal Sci.*, 98:208–218.

Eltayeb, I. A. (2023). Linear and nonlinear convective instabilities of a Maxwell - Cattaneo bidispersive layer in local thermal non-equilibrium. *J. Porous Media*, 26:55–81.

Eltayeb, I. A. and Elbashir, T. B. A. (2017). The stability of second sound waves in a rotating Darcy - Bénard porous layer in local thermal non-equilibrium. *Fluid Dynamics Research*, 49:045512.

Eltayeb, I. A., Hughes, D. W., and Proctor, M. R. E. (2020). The convective instability of a Maxwell - Cattaneo fluid in the presence of a vertical magnetic field. *Proc. Roy. Soc. London A*, 476:20200494.

Espinosa-Paredes, G. and Espinosa-Martinez, E. G. (2009). Fuel rod model based on non-Fourier heat conduction equation. *Annals of Nuclear Energy*, 36:680–693.

Fabrizio, M. (2011). Dario Graffi in a complex historical period. In *Mathematicians in Bologna 1861 – 1960*. Birkhauser.

Fabrizio, M., Franchi, F., and Nibbi, R. (2022). Second gradient Green - Naghdi type thermoelasticity and viscoelasticity. *Mech. Research Comm.*, 126:104014.

Fabrizio, M., Franchi, F., and Nibbi, R. (2023). Nonlocal continuum mechanics structures: the virtual powers method vs. the extra fluxes topic. *J. Thermal Stresses*, 46:75–87.

Fabrizio, M., Giorgi, C., and Morro, A. (2017). Modelling of heat conduction via fractional derivatives. *Heat and Mass Transfer*, 53:2785–2797.

Fabrizio, M. and Morro, A. (2003). *Electromagnetism of continuous media*. Oxford University Press, Oxford.

Falcón, N. (1998). Thermal instability for convection in astrophysical plasmas. *Astrophysics and Space Science*, 256:399–402.

Falcón, N. and Labrador, J. (2001). Thermal waves and unstable convection. *Odessa Astronomical Publications*, 14:141–143.

Famà, A., Restuccia, L., and Ván, P. (2021). Generalized ballistic - conductive heat transport laws in three - dimensional isotropic materials. *Continuum Mech. Thermodyn.*, 33:403–430.

Fan, Z., Zhao, J., Wang, D., Yue, F., Wu, Z., and Xue, R. (2024). Effects of local thermal non-equilibrium in hydraulic stimulation efficiency of enhanced geothermal systems. *Geomch. Geophys. Geo-energ. Geo-resour.*, 10:47.

Fernández, J. R. and Quintanilla, R. (2023). Uniqueness for a high order ill posed problem. *Proc. Roy. Soc. Edinburgh*, 153:1425–1438.

Ferreira, J. A. and de Oliveira, P. (2010). Looking for the lost memory in diffusion - reaction equations. In Rannacher, R. and Sequeira, A., editors, *Advances in Mathematical Fluid Mech.*, pages 229–251. Springer.

Fife, P. C. (1970). The Bénard problem for general fluid dynamical equations and remarks on the Boussinesq approximation. *Indiana Univ. Math. J.*, 20:303–326.

Fife, P. C. (2000). A gentle introduction to the physics and mathematics of incompressible flow. www.math.utah.edu/ fife/gentleb.pdf.

Fox, N. (1969). Low temperature effects and genralized thermoelasticity. *J. Inst. Maths. Applics.*, 5:373–386.

Franchi, F., Nibbi, R., and Straughan, B. (2017). Modelling bidispersive local thermal non-equilibrium flow. *Fluids*, 2:48.

Franchi, F., Nibbi, R., and Straughan, B. (2024). Sharp instability estimates for bidispersive convection with local thermal non-equilibrium. *Trans. Por. Media*, 151:193–211.

Franchi, F. and Straughan, B. (1994). Thermal convection at low temperature. *J. Non - Equilib. Thermodyn.*, 19:368–374.

Franchi, F. and Straughan, B. (2023). Dario Graffi e la sua influenza sulla Fisica Matematica. In Contucci, P., editor, *La matematica a Bologna dal dopoguerra*, pages 13–43. Bologna University Press. Mens Agitat - Colloquia - Accademia delle Scienze dell'Istituto di Bologna.

Freitas, R. B., Brandao, P. V., Alves, C. S. D. B., Celli, M., and Barletta, A. (2022). The effect of local thermal non - equilibrium on the onset of thermal instability for a metallic foam. *Phys. Fluids*, 34:034105.

Fried, E. and Gurtin, M. E. (2006). Tractions, balances, and boundary conditions for nonsimple materials with application to flow at small length scales. *Arch. Rational Mech. Anal.*, 182:513–554.

Fülöp, T., Kovács, R., Lovas, A., Rieth, A., Fodor, T., Szücs, M., Ván, P., and Gróf, G. (2018). Emergence of non - Fourier hierarchies. *Entropy*, 20:832.

Gaikwad, S. N. and Kamble, S. S. (2012). Analysis of kinear stability on double diffusion convection in a fluid saturated anisotropic porous layer with Soret effect. *Adv. Appl. Sci. Res.*, 3:1611–1617.

Gaikwad, S. N., Malashetty, M. S., and Rama Prasad, K. (2009). An analytical study of linear and nonlinear double diffusive convection in a fluid saturated anisotropic porous layer with Soret effect. *Appl. Math. Modelling*, 33:3617–3635.

Galdi, G. P. and Straughan, B. (1985a). Exchange of stabilities, symmetry, and nonlinear stability. *Arch. Rational Mech. Anal.*, 89:211–228.

Galdi, G. P. and Straughan, B. (1985b). A nonlinear analysis of the stabilizing effect of rotation in the Bénard problem. *Proc. Roy. Soc. London A*, 402:257–283.

Garaud, P. (2021). Journey to the centre of stars: the realm of low Prandtl number fluid dynamics. *Phys. Rev. Fluids*, 6:030501.

García Fernández, L., Bataller, H., Fruton, P., Giraudet, C., Vailati, A., and Croccolo, F. (2022). Stabilized convection in a ternary mixture with two Soret coefficients of opposite sign. *Eur. Phys. J. E*, 45:52.

Gentile, M. and Straughan, B. (2016). Hyperbolic diffusion with Christov - Morro theory. *Math. Comput. Simulation*, 127:94–100.

Gentile, M. and Straughan, B. (2017). Bidispersive vertical convection. *Proc. Roy. Soc. A*, 473:20170481.

Gentile, M. and Straughan, B. (2020). Bidispersive thermal convection with relatively large macropores. *J. Fluid Mech.*, 898:A14.

Gentile, M. and Straughan, B. (2024). Thermal convection with a Cattaneo heat flux model. *Proc. Roy. Soc. London A*, 480:20230771.

Gilbard, D. and Trudinger, N. S. (1977). *Elliptic partial differential equations of second order*, volume 224 of *Grundlehren der mathematischen Wissenschaften*. Springer, New York.

Giorgi, C., Morro, A., and Zullo, F. (2024). Modelling of heat conduction through rate equations. arXiv: 2402.00470 v1.

Giusteri, G. G., Marzocchi, A., and Musesti, A. (2011). Nonsimple isotropic incompressible linear fluids surrounding one - dimensional structures. *Acta Mech.*, 217:191–204.

Gossler, M. A., Bayer, P., and Zosseder, K. (2019). Experimental investigation of thermal retardation and local thermal non - equilibrium effects on heat transport in highly permeable, porous aquifers. *J. Hydrology*, 578:124097.

Govender, M. and Govinder, K. S. (2004). Generalized isothermal universes. *Int. J. Theor. Phys.*, 42:2253–2262.

Govender, M., Govinder, K. S., and Fleming, D. (2012). The role of pressure during shearing, dissipative collapse. *Int. J. Theor. Phys.*, 51:3399–3409.

Govender, M. and Thirukkanesh, S. (2009). Dissipative collapse in the presence of λ. *Int. J. Theor. Phys.*, 48:3558–3566.

Graffi, D. (1936). Sopra alcuni fenomeni ereditari dell'elettrologia. *Rend. Ist. Lomb. Sc. Lett.*, 19:151–166.

Graffi, D. (1963). Sulla propagazione nei mezzi dispersivi. *Annali di Matematica Pura ed Applicata*, 60:173–193.

Graffi, D. (1965). Su un'equazione integro - differenziale di tipo iperbolico. In *Seminari dell'Istituto Nazionale di Alta Matematica 1962-63*, pages 123–132, Roma. Cremonese.

Grifò, G., Consolo, G., Currò, C., and Valenti, G. (2023). Rhombic and hexagonal pattern formation in 2D hyperbolic reaction - transport systems in the context of dryland ecology. *Physica D*, 449:133745.

Hema, M., Shivakumara, I. S., and Ravisha, M. (2020). Double diffusive LTNE porous convection with Cattaneo effects in the solid. *Heat Transfer*, 49:3613–3629.

Hema, M., Shivakumara, I. S., and Ravisha, M. (2021). Cattaneo - LTNE effects on the stability of Brinkman convection in an anisotropic porous layer. *Int. J. Appl. Comput. Math.*, 7:38.

Herrera, L. (2019). Causal heat conduction contravening the fading memory paradigm. *Entropy*, 21:950.

Herrera, L. and Falcón, N. (1995a). Convection theory before relaxation. *Astrophysics and Space Science*, 234:139–152.

Herrera, L. and Falcón, N. (1995b). Heat waves and thermohaline instability in a fluid. *Physics Letters A*, 201:33–37.

Herrera, L. and Falcón, N. (1995c). Secular stability behaviour of nuclear burning before relaxation. *Astrophysics and Space Science*, 229:105–115.

Herrera, L. and Martínez, J. (1998). Dissipative fluids out of hydrpstatic equilibrium. *Class. Quantum Grav.*, 15:407–420.

Herrera, L. and Santos, N. O. (1997). Thermal evolution of compact objects and relaxation time. *Mon. Not. R. Astron. Soc.*, 287:161–164.

Herrera, L. and Santos, N. O. (2004). Dynamics of dissipative gravitational collapse. *Phys. Rev. D*, 70:084004.

Hughes, D. W., Proctor, M. R. E., and Eltayeb, I. A. (2021). Maxwell - Cattaneo double diffusive convection: limiting cases. *J. Fluid Mech.*, 927:A13.

Hughes, D. W., Proctor, M. R. E., and Eltayeb, I. A. (2022). Rapidly rotating Maxwell - Cattaneo convection. *Phys. Rev. Fluids*, 7:093502.

Iesan, D. (2023). Thermal stresses that depend on temperature gradients. *ZAMP*, 74:138.

Jia, B. and Jian, Y. (2024). Effect of Oldroyd - B fluid on thermal convection cwstability in a bidispersive porous medium with relatively large macropores. *Phys. Fluids*, 36:014105.

Jordan, P. M. (2005). Growth and decay of shock and acceleration waves in a traffic flow model with relaxation. *Physica D*, 207:220–229.

Jordan, P. M. (2015). Second sound propagation in rigid, nonlinear conductors. *Mech. Res. Communications*, 68:52–59.

Jordan, P. M., Dai, W., and Mickens, R. E. (2008). A note on the delayed heat equation; instability with respect to initial data. *Mech. Res. Comm.*, 35:414–420.

Jordan, P. M. and Lambers, J. V. (2021a). On the propagation and bifurcation of singular surface shocks under a class of wave equations based on second sound and flux models and logisitic growth. *Int. J. Nonlinear Mech.*, 132:103696.

Jordan, P. M. and Lambers, J. V. (2021b). Revisiting manne et al. (2000); a reformulation and alternative interpretation under the modified internal energy theory of second sound. *Wave Motion*, 126:102756.

Jou, D., Casas Vázquez, J., and Lebon, G. (2010). *Extended irreversible thermodynamics*. Springer, New York, fourth edition.

Kalantarov, V. K., Levant, B., and Titi, E. S. (2009). Gevrey regularity of the global attractor of the 3D Navier - Stokes - Voigt equations. *J. Nonlinear Sci.*, 19:133–152.

Kalantarov, V. K. and Titi, E. S. (2009). Global attractors and determining modes for the 3D Navier - Stokes - Voigt equations. *Chinese Annals of Math.*, 30:697–714. Ser. B.

Kaltenbacher, B., Lasiecka, I., and Pospieszalska, M. K. (2012). Well - posedness and exponential decay of the energy in the nonlinear Jordan - Moore - Gibson - Thompson equation arising in high intensity ultrasound. *Math. Models Meth. Appl. Sci.*, 22:1250035.

Kaltenbacher, B. and Nikolić, V. (2019). The Jordan - Moore - Gibson - Thompson equation: well - posedness with quadratic gradient nonlinearity and singular limit for vanishing relaxation time. *Math. Models Meth. Appl. Sci.*, 29:2523–2556.

Kaltenbacher, B. and Nikolić, V. (2021). The inviscid limit of third - order linear and nonlinear acoustic equations. arXiv: 2101.05488v2.

Käpylä, P. J. (2021). Prandtl number dependence of stellar convection: flow statistics and convective energy transport. *Astronomy and Astrophysics*, 655:A78.

King, A. C., Needham, D. J., and Scott, N. H. (1998). The effects of weak hyperbolicity on the diffusion of heat. *Proc. Roy. Soc. London A*, 454:1659–1679.

Kovács, R. and Ván, P. (2015). Generalized heat conduction in heat pulse experiments. *Int. J. Heat Mass Transfer*, 18:613–620.

Kovács, R. and Ván, P. (2018). Second sound and ballistic heat conduction: NaF experiments revisited. *Int. J. Heat Mass Transfer*, 117:682–690.

Lakshmi, K. M., Laroze, D., and Siddheshwar, P. G. (2022). Natural convection of a binary liquid in cylindrical porous annuli / rectangular porous enclosures with cross - diffusion effects under local thermal nonequilibrium state. *Int. J. Heat Mass Transfer*, 184:122294.

Larmor, J. and Thomson, J. (1912). On a changing tesselated structure in certain liquids (due to convective circulation). In *Collected papers in Physics and Engineering, by James Thomson*, pages 136–141. Cambridge University Press.

Layton, W. J. and Rebholz, L. G. (2013). On relaxation times in the Navier - Stokes - Voigt model. *Int. J. Comp. Fluid Dyn.*, 27:184–187.

Liu, H., Bussmann, M., and Mostaghimi, J. (2009). A comparison of hyperbolic and parabolic models of phase change of a pure metal. *Int. J. Heat Mass Transfer*, 52:1177–1184.

López Molina, J. A., Rivera, M. J., Trujillo, M., and Berjano, E. J. (2008). Effect of the thermal wave in radiofrequency ablation modelling: an analytical study. *Phys. Med. Biol.*, 53:1447–1462.

Lu, W. S. and Shao, H. Y. (2003). Generalized nonlinear subcritical symmetric instability. *Adv. Atmos. Sci.*, 20:623–630.

Mahfoudh, I., Principi, P., Fioretti, R., and Safi, M. (2019). Experimental studies on the effect of using phase change materials in a salinity - gradient solar pond under a solar simulator. *Solar Energy*, 186:335–346.

Mamatha, A. L., Ravisha, M., and Shivakumara, I. S. (2022). Chaotic Cattaneo - LTNE porous convection. *Waves in Random and Complex Media*, pages 1–23. https://doi.org/10.1080/17455030.2022.2155320.

Marchand, R., McDevitt, T., and Triggiani, R. (2012). An abstract semigroup approach to the third - order Moore - Gibson - Thompspn partial differential equation arising in high intensity ultrasound: structural decomposition, spectral analysis, exponential stability. *Math. Meth. Appl. Sci.*, 35:1896–1929.

Mariano, P. M. (2017). Finite - speed heat propagation as a consequence of microstructural changes. *Continuum Mechanics and Thermodynamics*, 29:1241–1248.

Mariano, P. M. (2023). Proof of Straughan's claim on Payne - Song's and modified Guyer - Krumhansl equations. *Proc. Roy. Soc. London A*, 479:20230439.

Mariano, P. M., Polikarpus, J., and Spadini, M. (2022). Solutions of linear and nonlinear schemes for non - Fourier heat conduction. *Int. J. Heat Mass Transfer*, 183:122193.

Markowitz, H. (1977). Boltzmann and the beginnings of linear viscoelasticity. *Trans. Soc. Rheology*, 21:381–398.

Matveeva, O. P. (2013). Model of thermoconvection of incompressible viscoelastic fluid of non - zero order - computational experiment. *Bull. South Ural State Tech. Univ., Ser. Math. Modelling and Programming*, 6:134–138.

Maxwell, J. C. (1867). On the dynamical theory of gases. *Phil. Trans. Roy. Soc. London*, 157:49–88.

Meyer, R. J. (2006). Ultrasonic drying of saturated porous solids via second sound. http://www.freepatentsonline.com/6376145.html.

Mihaljan, J. (1962). A rigorous exposition of the Boussinesq approximation applicable to a thin layer of fluid. *Astrophys. J.*, 136:1126–1133.

Miller, R. K. (1978). An integrodifferential equation for rigid heat conductors with memory. *J. Math. Anal. Appl.*, 66:313–332.

Miranville, A. and Quintanilla, R. (2009). A generalization of the Caginalp phase - field system based on the Cattaneo law. *Nonlinear Analysis; Theory, Methods and Applications*, 71:2278–2290.

Mitra, K., Kumar, S., Vedavarz, A., and Moallemi, M. K. (1995). Experimental evidence of hyperbolic heat conduction in processed meat. *Trans. ASME, J. Heat Transfer*, 117:568–573.

Morro, A. (2017a). A thermodynamic approach to rate equations in Continuum Physics. *J. Physical Sci. Applications*, 7:15–23.

Morro, A. (2017b). Thermodynamic consistency of objective rate equations. *Mech. Res. Communications*, 84:72–76.

Morro, A. (2018). Modelling elastic heat conductors via objective rate equations. *Cont. Mech. Thermodyn.*, 30:1231–1243.

Morro, A. (2022). Objective equations of heat conduction in deformable bodies. *Mech. Res. Communications*, 125:103979.

Morro, A. (2024). On the modelling of thermal convection in porous media through rate - type equations. *Annali dell'Università di Ferrara*, 70:1–17.

Morro, A. and Giorgi, C. (2020). Objective rate equations and memory properties in continuum physics. *Math. Comp. Simulation*, 176:243–253.

Morro, A. and Giorgi, C. (2023). *Mathematical modelling of continuum physics*. Modeling and Simulation in Science, Engineering and Technology. Birkhauser.

Musesti, A. (2009). Isotropic linear constitutive relations for nonsimple fluids. *Acta Mech.*, 204:81–88.

Nandal, R., Siddheshwar, P. G., and Neela, D. (2023). Study of influence of combustion on Darcy - Brinkman convection with inherent local thermal nonequilibrium between phases. *Transport in Porous Media*, 146:741–769.

Nield, D. A. and Bejan, A. (2013). *Convection in porous media*. Springer, New York, fourth edition.

Nield, D. A. and Kuznetsov, A. V. (2006). The onset of convection in a bidispersive porous medium. *Int. J. Heat Mass Transfer*, 49:3068–3076.

Nika, G. (2023). A gradient system for a higher gradient generaslization of Fourier's law of heat conduction. *Modern Physics Letters B*, 37:2350011.

Nika, G. and Muntean, A. (2022). Hypertemperature effects in heterogeneous media and thermal flux at small length scales. *Networks and Heterogeneous Media*, 12:1207–1225.

Nikolić, V. and Said Houari, B. (2021). On the Jorden - Moore - Gibson - Thompson wave equation in hereditary fluids with quadratic gradient nonlinearity. *J. Math. Fluid Mech.*, 23:3.

Niwa, H. (1998). Migration dynamics of fish schools in heterothermal environments. *J. Theor. Biol.*, 193:215–231.

Oskolkov, A. P. (1988). Initial - boundary value problems for the equations of Kelvin - Voigt fluids and Oldroyd fluids. *Proc. Steklov Inst. Math.*, 179:126–164.

Oskolkov, A. P. (1995). Nonlocal problems for the equations of motion of Kelvin - Voigt fluids. *J. Math. Sciences*, 75:2058–2078.

Oskolkov, A. P. and Shadiev, R. (1994). Towards a theory of global solvability on $[0, \infty)$ of initial - boundary value problems for the equations of motion of Oldroyd and Kelvin - Voigt fluids. *J. Math. Sciences*, 68:240–253.

Papanicolaou, N. C., Christov, C. I., and Jordan, P. M. (2011). The influence of thermal relaxation on the oscillatory properties of two - gradient convection in a vertical slot. *Euro. J. Mech. - B/Fluids*, 30:68–75.

Payne, L. E. (1964). Uniqueness criteria for steady state solutions of the Navier - Stokes equations. In *Atti del Simposio Internazionale sulle Applicazioni dell'Analisi alla Fisica Matematica, Cagliari - Sassari, 28-IX to 4-X, 1964*, pages 130–153, Roma. Edizioni Cremonese.

Payne, L. E. and Song, J. C. (1997). Continuous dependence on initial - time geometry and spatial geometry in generalized heat conduction. *J. Math. Anal. Appl.*, 214:173–190.

Platten, J. K. (2006). The Soret effect: a review of recent experimental results. *J. Appl. Mech.*, 73:5–15.

Quintanilla, R. (2008). A well - posed problem for the dual - phase - lag heat conduction. *J. Thermal Stresses*, 31:260–269.

Quintanilla, R. and Jordan, P. M. (2009). A note on the two temperature theory with dual-phase-lag delay: Some exact solutions. *Mechanics Research Communications*, 36:796–803.

Quintanilla, R. and Racke, R. (2007). Qualitative aspects in dual-phase-lag heat conduction. *Proc. Roy. Soc. London A*, 463:659–674.

Racke, R. and Said Houari, B. (2021). Global well-posedness of the Cauchy problem for the 3D Jordan - Moore - Gibson - Thompson equation. *Comm. Contemporary Math.*, 23:2050069.

Rees, D. A. S. (2003). Vertical free convective boundary - layer flow in a porous medium using a thermal nonequilibrium model: elliptical effects. *ZAMP*, 54:437–448.

Rees, D. A. S. (2009). Microscopic modelling of the two - temperature model for conduction in heterogeneous three - dimensional media. In *Proc. 4th Int. Conf. on Applications of Porous Media, Istanbul.*

Rees, D. A. S. (2010). Microscopic modelling of the two - temperature model for conduction in heterogeneous media. *J. Porous Media*, 13:125–143.

Rees, D. A. S. (2011). The effect of local thermal nonequilibrium on the stability of convection in a vertical porous channel. *Trans. Porous Media*, 87:459–464.

Rees, D. A. S. and Bassom, A. P. (2010). The radial injection of a hot fluid into a cold porous medium: rhe effects of local thermal non-equilibrium. *Comput. Therm. Science*, 2:221–230.

Rees, D. A. S., Bassom, A. P., and Siddheshwar, P. G. (2008). Local thermal non-equilibrium effects arising from the injection of a hot fluid into a porous medium. *J. Fluid Mech.*, 594:379–398.

Rees, D. A. S. and Pop, I. (2010). Vertical free convective boundary - layer flow in a porous medium using a thermal nonequilibrium model. *J. Porous Media*, 2:31–44.

Rodrigues, J. F. and Urbano, J. M. (1998). On the stationary Boussinesq - Stefan problem with constitutive power laws. *Int. J. Nonlinear Mech.*, 33:555–566.

Rogolino, P. and Cimmelli, V. A. (2021). Differential consequences of balance laws in extended irreversible thermodynamics of rigid heat conductors. *Proc. Roy. Soc. London A*, 475:20180482.

Ruggeri, T. (1983). Symmetric - hyperbolic system of conservative equations for a viscous heating conducting fluid. *Acta Mech.*, 47:167–183.

Ruggeri, T. and Sugiyama, M. (2021). *Classical and relativistic rational extended thermodynamics of gases*. Springer, Cham, Switzerland.

Said Houari, B. (2022). Global well-posedness of the Cauchy problem for the Jordan - Moore - Gibson - Thompson equation with arbitrarily large high - order Sobolev norms. *Discrete and Continuous Dynamical Sustems*, 42:4615–4635.

Saravanan, S. and Vigneshwaran, S. (2020). Centrifugal filtration convection in bidisperse media. *Phys. Fluids*, 32:084109.

Sellitto, A., Cimmelli, V. A., and Jou, D. (2013). Entropy flux and anomalous axial heat transport at the nanoscale. *Phys. Rev. B*, 87:054302.

Sellitto, A., Zampoli, V., and Jordan, P. M. (2020). Second sound beyond Maxwell - Cattaneo: nonlocal effects in hyperbolic heat transfer at the nanoscale. *Int. J. Engng. Sci.*, 154:103328.

Shevtsova, V. M., Melnikov, D. E., and Legros, J. C. (2006). Onset of convection in Soret driven instability. *Phys. Rev. E*, 73:047302.

Shivakumara, I. S. and Raghunatha, A. (2022). Changes in the onset of double - diffusive local thermal nonequilibrium convection due to the introduction of a third component. *Transport in Porous Media*, 143:225–242.

Shivakumara, I. S., Raghunatha, K. R., Dhananjaya, M., and Vinod, Y. (2023). Lack of local thermal nonequilibrium effects on convection in a porous medium saturated with an Ellis fluid. *Transport Por. Media*, 146:703–720.

Shivakumara, I. S., Ravisha, M., Ng, C. O., and Varun, V. L. (2015a). Porous ferroconvection with local thermal nonequilibrium temperatures and with Cattaneo effects in the solid. *Acta Mech.*, 226:3763–3779.

Shivakumara, I. S., Ravisha, M., Ng, C. O., and Varun, V. L. (2015b). A thermal nonequilibrium model with Cattaneo effects for convection in a Brinkman porous layer. *Int. J. Non-linear Mech.*, 71:39–47.

Shyabal, S., Ravisha, M., Hanumagowda, B. N., Mamatha, A. L., and Shivakumara, I. S. (2023). Onset of LTNE anisotropic porous convection: effect of asymmetric temperature boundary conditions. *European Phys. J. Plus*, 138:106.

Siddabasappa, C., Siddheshwar, P. G., and Mallikarjunaiah, S. N. (2023). Analytical study of Brinkman - Bénard convection in a bidisperse porous medium: Linear and weakly nonlinear study. *Thermal Science and Engineering Progress*, 39:101696.

Siddheshwar, P. G., Sakshath, T. N., and Siddabasappa, C. (2021). Effect of rotation on Brinkman - Bénard convection of a Newtonian nanoliquid using local thermal nonequilibrium model. *Thermal Science and Engineering Progress*, 25:100994.

Singh, M., Ragoju, R., Reddy, G. S. K., Matta, A., Paidipati, K. K., and Chesneau, C. (2022). Non-linear magnetoconvection in a bidispersive porous layer: a Brinkman model. *Earth Science Informatics*, 15:2171–2180.

Sneddon, I. N. (1995). *Fourier transforms*. Dover.

Soret, C. (1879). Sur l'état d'équilibre que prend au point de vue de sa concentration une dissolution saline primitivement homohéne dont deux parties sont portées à des températures différentes. *Arch. Sci. Phys. Nat.*, 2:48–61.

Spiegel, E. and Veronis, G. (1960). On the Boussinesq approximation for a compressible fluid. *Astrophys. J.*, 131:442–447.

Straughan, B. (2000). A sharp nonlinear stability threshold in rotating porous convection. *Proc. Roy. Soc. London A*, 457:87–93.

Straughan, B. (2004). *The energy method, stability, and nonlinear convection*, volume 91 of *Appl. Math. Sci.* Springer, New York, second edition.

Straughan, B. (2006). Global nonlinear stability in porous convection with a thermal non-equilibrium model. *Proc. Roy. Soc. London A*, 462:409–418.

Straughan, B. (2007). A note on convection with nonlinear heat flux. *Ricerche di Matematica*, 56:229–239.

Straughan, B. (2010a). Porous convection with Cattaneo heat flux. *Int. J. Heat Mass Transfer*, 53:2808–2812.

Straughan, B. (2010b). Thermal convection with the Cattaneo - Christov model. *Int. J. Heat Mass Transfer*, 53:95–98.

Straughan, B. (2011a). *Heat waves*, volume 177 of *Appl. Math. Sci.* Springer, New York.

Straughan, B. (2011b). Tipping points in Cattaneo - Christov thermohaline convection. *Proc. Roy. Soc. London A*, 467:7–18.

Straughan, B. (2013). Porous convection with local thermal non - equilibrium temperatures and with Cattaneo effects in the solid. *Proc. Roy. Soc. London A*, 469:20130187.

Straughan, B. (2015a). *Convection with local thermal non-equilibrium and microfluidic effects*, volume 32 of *Advances in Mechanics and Mathematics Series*. Springer, Cham, Switzerland.

Straughan, B. (2015b). Exchange of stability in Cattaneo - LTNE porous convection. *Int. J. Heat Mass Transfer*, 89:792–798.

Straughan, B. (2016). Importance of Darcy or Brinkman laws upon resonance in thermal convection. *Ricerche Matem.*, 65:349–362.

Straughan, B. (2018). Horizontally isotropic bidispersive thermal convection. *Proc. Roy. Soc. London A*, 474:20180018.

Straughan, B. (2019). Anisotropic bidispersive convection. *Proc. Roy. Soc. London A*, 475:20190206.

Straughan, B. (2021a). Competitive double diffusive convection in a Kelvin - Voigt fluid of order one. *Appl. Math. Optimization*, 84:631–650.

Straughan, B. (2021b). Thermosolutal convection with a Navier - Stokes - Voigt fluid. *Appl. Math. Optimization*, 83:2587–2599.

Straughan, B. (2022a). Continuous dependence and convergence for a Kelvin - Voigt fluid of order one. *Annali Univ. Ferrara*, 68:49–61.

Straughan, B. (2022b). Instability thresholds for thermal convection in a Kelvin - Voigt fluid of variable order. *Rend. Circ. Matem. Palermo Ser. 2*, 71:187–206.

Straughan, B. (2023a). Nonlinear stability for convection with temperature dependent viscosity in a Navier - Stokes - Voigt fluid. *European Physical Journal Plus*, 138:438.

Straughan, B. (2023b). Rotating convection in a higher gradient Navier - Stokes fluid. *European Physical Journal Plus*, 138:640.

Straughan, B. (2023c). Thermal convection in a Brinkman - Darcy - Kelvin - Voigt fluid with a generalized Maxwell - Cattaneo law. *Annali dell'Universitá di Ferrara*, 69:521–540.

Straughan, B. (2023d). Thermal convection in a higher gradient Navier - Stokes fluid. *European Physical Journal Plus*, 138:60.

Straughan, B. and Franchi, F. (1984). Bénard convection and the Cattaneo law of heat conduction. *Proc. Roy. Soc. Edinburgh A*, 96:175–178.

Stupelis, L. (2000). An initial - boundary value problem for a system of equations of magnetohydrodynamics. *Lithuanian Mathematical J.*, 40:176–196.

Stupelis, L. (2004a). Investigation of convection flows in an electromagnetic field. *Lithuanian Mathematical J.*, 44:395–436.

Stupelis, L. (2004b). The MHD equations in the case of low conductivity. *Nonlinear Anal. Real World Applicns.*, 5:1–32.

Stupelis, L. (2007). On solvability of an initial - boundary - value problem for equations of magnetohydrodynamics. *Lithuanian Mathematical J.*, 47:195–227.

Sukacheva, T. G. and Kondyukov, A. O. (2014). On a class of Sobolev type equations. *Bull. South Ural State Tech. Univ., Ser. Math. Modelling and Programming*, 7:5–21.

Sukacheva, T. G. and Matveeva, O. P. (2010). On a homogeneous model of the non - compressible viscoelastic Kelvin - Voigt fluid of the non - zero order. *J. Samara State Tech. Univ., Ser. Phys. Math. Sci.*, 5:33–41.

Thompson, J. T., Taylor, K. R., and Gentile, C. (2010). Gradients of strain and strain rate in the hollow muscular organs of soft - bodied animals. *Biol. Lett.*, 6:482–485.

Thomson, J. (1882). On a changing tesselated structure in certain liquids. *Proc. Phil. Soc. Glasgow*, 13:464–468.

Toy, V., Benson, P., Castro, J., Doan, M. L., De Siena, L., Enzmann, F., and Tajcmanova, L. (2019). Dual porosity systems (pore and fracture) permeability in volcanic tuffs by computed tomography, Cimino - Vico volcanoes (Italy). Terrestial Magmatic Systems (temas.uni-mainz.de).

Tripathi, V. K., Shankar, B. M., Mahajan, A., and Shivakumara, I. S. (2024). Global nonlinear stability of bidispersive porous convection with throughflow and depth - dependent viscosity. *Phys. Fluids*, 36:014110.

Truesdell, C. (1955). The simplest rate theory of pure elasticity. *Comm. Pure Appl. Math.*, 8:123–132.

Truesdell, C. and Muncaster, R. G. (1980). *Fundamentals of Maxwell's kinetic theory of a simple monatomic gas*. Academic Press, New York.

Vadasz, J. J., Govender, S., and Vadasz, P. (2005). Heat transfer enhancement in nano-fluids suspensions: possible mechanisms and explanations. *Int. J. Heat Mass Transfer*, 48:2673–2683.

Vadasz, P. (1998). Coriolis effect on gravity - driven convection in a rotating porous layer heated from below. *J. Fluid Mech.*, 376:351–375.

Ván, P., Berezovski, A., Fülöp, T., Gróf, G., Kovács, R., Lovas, A., and Verhás, J. (2017). Guyer - Krumhansl heat conduction at room temperature. *EPL*, 118:50005.

Ván, P. and Fülöp, T. (2012). Universality in heat conduction theory: weakly nonlocal thermodynamics. *Annalen der Physik*, 524:470–478.

Vedavarz, A., Mitra, K., Kumar, S., and Moallemi, M. K. (1992). Effect of hyperbolic heat conduction on temperature distribution in laser irradiated tissue with blood perfusion. *Adv. Biological Heat Mass Transfer, ASME HTD*, 231:7–16.

Vieira, L. D., Moreira, A. C., Mantovani, I. F., Honorato, A. R., Prado, O. F., Becker, M., Fernandes, C. P., and Waichel, B. L. (2021). The influence of secondary processes on the porosity of volcanic rocks: a multiscale analysis using 3D X-ray microtomography. *Appl. Radiation and Isotopes*, 172:109657.

Wang, H., Zhang, L. G., and Mei, Y. Y. (2018). Investigation of the exergy performance of salt gradient solar ponds with porous media. *Int. J. Exergy*, 25:34–53.

Wang, Z. and Hillen, T. (2008). Shock formation in a chemotaxis model. *Math. Meth. Appl. Sci.*, 31:45–70.

Weber, E. H. (1855). Mikroskopische Beobachtungen sehr gesetz mässiger Bewegungen: welche die Bildung von Niederchlägen harziger Körper aus Weingeist begleiten. *Ann. Physik*, 170:447–459.

Wesfreid, J. E. (2017). Henri Bénard; thermal convection and vortex shedding. *Comptes Rendus Mecanique*, 345:446–466.

Zampoli, V. and Jordan, P. M. (2022). Second - sound phenomena in type II conductors with Stefan - Boltzmann source. *Mech. Res. Comm.*, 126:103998.

Zampoli, V. and Jordan, P. M. (2024). Front propagation under the linearized Eringen - Cattaneo - Christov - Straughan model. *J. Non-Equilibrium Thermodynamics*, 49:237–249.

Index

B. Straughan, *Thermal Convection with a Cattaneo Flux Law*, Advances in Mechanics and Mathematics 53, https://doi.org/10.1007/978-3-031-80885-2

SPRINGER NATURE

GPSR Compliance

The European Union's (EU) General Product Safety Regulation (GPSR) is a set of rules that requires consumer products to be safe and our obligations to ensure this.

If you have any concerns about our products, you can contact us on ProductSafety@springernature.com

In case Publisher is established outside the EU, the EU authorized representative is:

Springer Nature Customer Service Center GmbH
Europaplatz 3
69115 Heidelberg, Germany

The manufacturer's authorised representative in the EU is Springer Nature Customer Service Centre GmbH, Europaplatz 3, 69115 Heidelberg, Germany. If you have any concerns regarding our products, please contact ProductSafety@springernature.com

Printed and bound by CPI Group (UK) Ltd, Croydon, CR0 4YY
01/10/2026
02225876-0018